Solvents Theory and Practice

Solvents Theory and Practice

Roy W. Tess, *Editor*

A symposium sponsored by the Division of Organic Coatings and Plastics Chemistry at the 162nd Meeting of the American Chemical Society, Washington, D.C., Sept. 15-16, 1971.

ADVANCES IN CHEMISTRY SERIES **124**

AMERICAN CHEMICAL SOCIETY

WASHINGTON, D. C.   1973

ADCSAJ 124 1-227 (1973)

Copyright © 1973

American Chemical Society

All Rights Reserved

Library of Congress Catalog Card 73-88797

ISBN 8412-0186-2

PRINTED IN THE UNITED STATES OF AMERICA

Advances in Chemistry Series
Robert F. Gould, *Editor*

Advisory Board

Bernard D. Blaustein

Paul N. Craig

Ellis K. Fields

Edith M. Flanigen

Egon Matijević

Thomas J. Murphy

Robert W. Parry

Aaron A. Rosen

Charles N. Satterfield

FOREWORD

ADVANCES IN CHEMISTRY SERIES was founded in 1949 by the American Chemical Society as an outlet for symposia and collections of data in special areas of topical interest that could not be accommodated in the Society's journals. It provides a medium for symposia that would otherwise be fragmented, their papers distributed among several journals or not published at all. Papers are refereed critically according to ACS editorial standards and receive the careful attention and processing characteristic of ACS publications. Papers published in ADVANCES IN CHEMISTRY SERIES are original contributions not published elsewhere in whole or major part and include reports of research as well as reviews since symposia may embrace both types of presentation.

CONTENTS

Preface		ix
1.	**Trends in Solvent Science and Technology** Harry Burrell	1
2.	**Predicted Compositions during Mixed Solvent Evaporation from Resin Solutions Using the Analytical Solutions of Groups Method** E. L. Derr and C. H. Deal	11
3.	**Calculation of Concentrated Polymer Solution Viscosities: A New Approach** M. J. Hillyer and W. J. Leonard	31
4.	**Solvent Selection by Computer** Charles M. Hansen	48
5.	**Prediction of Flash Points for Solvent Mixtures** John G. Walsham	56
6.	**The Photochemical Smog Reactivity of Organic Solvents** Arthur Levy	70
7.	**Photochemical Smog and the Atmospheric Reactions of Solvents** John L. Laity, Israel G. Burstain, and Bruce R. Appel	95
8.	**A Practical Approach to Solvent Applications in Coatings and Inks** D. K. Sausaman	113
9.	**Solvent Systems for Hydrocarbon Resins** Paul O. Powers	131
10.	**Solvents for Use in Electrodeposition Coatings** C. A. May	141
11.	**Polyamide Resin Solubility Parameters** E. R. Hinden and P. D. Whyzmuzis	168
12.	**Technique for Reformulating Solvent Mixtures in Epoxy Resin Coatings** George R. Somerville and John A. Lopez	175
13.	**Solubility Characteristics of Today's Vinyl Chloride Homopolymers, Copolymers, and Terpolymers** Russell A. Park	186
Index		219

PREFACE

The term solvents, as used here, refers to organic liquids used to dissolve resins and polymers principally for application in surface coatings. About 3.5 billion pounds of solvents with a value of about 275 million dollars are used annually in the United States for this purpose.

Over the past several decades, knowledge of the properties of solvents and resin solutions has been acquired by laborious experimental work. Especially important properties are the evaporation characteristics and solvent power of neat solvents and the viscosity of resin solutions.

As the technology of solvents and solutions progressed, interest increased in the theoretical basis for the observed properties. Although the concept of solubility parameter was described many years ago by Hildebrand, interest in this concept was not great judging by the limited work on the topic. It was about 1955 before more active interest was generated in solubility parameter and other aspects of the physical chemistry of solvents. Harry Burrell was instrumental in stimulating the interest of the coatings industry in solubility parameter in 1955 with his noted paper in the *Official Digest* (now *Journal of Paint Technology*). Especially since then, many academic and industrial scientists have made important contributions to the theoretical knowledge of solvents. At the heart of the matter was the need and desire to describe solvents and their blends in fundamental terms which deal with the forces among solvent and dissolved resin molecules and the correlation of these fundamental properties to practical properties like viscosity of resin solutions and evaporation phenomena. The Paint Research Institute has been a catalyst in stimulating advances in the field by sponsoring productive fellowships at leading universities. Much of this work, and other contributions, have been published in the *Journal of Paint Technology* since 1965. A short course on solubility parameters was sponsored by the Paint Industry Education Bureau and presented by Kent State University in July 1972.

Investigations of solvents received tremendous impetus in the late sixties with the concern to control air pollution. The well-known Rule 66 of Los Angeles County became effective in 1967 for large solvent users; it classified solvents according to their propensity to induce photochemical smog as manifested principally by eye irritation, but also by oxidant formation and other effects. The adoption and proliferation of air pollution regulations generated the need for industry to reformulate thousands of solvent formulations to comply with the composition limitations of the air pollution laws. The computer was enlisted to facilitate the reformulation effort; its effectiveness depended upon the foundation of workable theoretical understanding of the physical chemistry of solvents and solutions. The computer could calculate rapidly a selection of worthy candidate solvent systems which could be checked in the laboratory. Suggested computer formulas met the composition limitations of air pollution laws as well as constraints of evaporation rate, viscosity, and the various solution parameters.

Rapid changes have also occurred in resin compositions and in application techniques. These advances have created a demand for new knowledge of solvent systems including combinations of organic solvents with water.

This volume covers some new ideas and results along with a modest amount of review in the theory of solvents and resin solutions. Two comprehensive papers cover recent and new information on the pertinent chemistry of solvents as related to air pollution. Finally, several reports are concerned with approaches used by industry to select solvent systems and with the actual solvent systems suitable for some prominent resin types. Solvents used in surface coatings for electrodeposition are covered in an extensive paper.

Although comprehensive coverage of solvents would require several books, this volume covers some areas of current interest. It is hoped that these reports will aid and stimulate scientists and practitioners in their study and use of solvents for coatings and related products.

Roy W. Tess

Shell Chemical Co.
Houston, Texas
July 1973

Trends in Solvent Science and Technology

HARRY BURRELL

Inmont Corp., Central Research Lab, 1255 Broad St., Clifton, N. J. 07015

The solubility parameter concept has been extremely useful in solving coating formulation problems. The theory was adapted for practical uses when the importance of hydrogen bonding was recognized and when data were reported in terms of a solubility parameter range. The concept has also been applied to polymer swelling, distillation and extraction, and solubility of gases. Much work is now going on to quantify hydrogen bonding and polar factors, as well as applications to pigment dispersion and interfacial energy studies. The limitations imposed by entropy considerations and Prigogine theory have been discussed. Other theoretical approaches being explored are Lewis acid-base concepts and the domain theory of molecular clustering. The future role of solvents in coatings technology must be seriously examined because of ecological factors long ignored. Usage will be limited by the success of aqueous systems, preformed films, high solids liquid, and powdered coatings.

Probably the most useful theory of solubility is that proposed by Hildebrand (1). This is based on the thermodynamically sound principle that the free energy change upon dissolution of a polymer in a solvent must be negative. The free energy change can be calculated from the equation

$$\Delta F = \Delta H - T\Delta S$$

Many of the polymers used in the coatings and plastics industries are highly amorphous. The lack of crystallinity in such polymers allows their solubility behavior to be treated as a simple mixing procedure. The entropy of a liquid is relatively large because the molecules are free to move about within the confines of a container as well as to rotate and align themselves with neighboring molecules in a random manner; depending on the vapor pressure they may even move through the liquid surface and enter the gas phase. A polymer molecule on the other hand, because of its size and shape, does not have much freedom of translation past its neighboring molecules; it is also so thoroughly entangled with its neighbors that the freedom of rotation and even of segmental motion is inhibited. When small solvent molecules are mixed with large polymer molecules, the latter accept a great increase in entropy. They are now more free to move about as a whole (depending on viscosity), but in addition their configurational entropy is greatly increased because the newly obtained freedom of rotation about carbon–to–carbon bonds allows a molecule to contort into many different shapes. The ΔS term in the above equation is therefore relatively large, and since the term is negative, the product of temperature times entropy tends to make the free energy negative. The restricting factor therefore is the ΔH term which measures the heat of mixing.

In a mixing process, the molecules of each component are separated by the interposition of a molecule from the second component. Hildebrand reasoned that the energy of this separation (to an infinite distance) could be measured by the heat of vaporization. If the energy to separate the molecules of component A is sufficiently different from the energy required to separate the molecules of component B, each species of molecule will prefer its like neighbors and will refuse to mix with the molecules of the other component. On the other hand, if the energies required to separate these two species are relatively similar, the molecules of component A will tolerate the proximity of B molecules and vice-versa. Hildebrand, therefore, proposed that the square root of the energy, in calories per cc, required to separate the molecules of a given substance be designated as solubility parameter, δ. If the solubility parameters of two substances were identical, then they would be miscible in all proportions. Because the entropy term in the free energy equation has a finite negative value, there would be an allowable difference in solubility parameters which would still permit mixing.

When this theory was used to predict the solubility of polymers in a variety of solvents, it was only partially successful. It was apparent that other intermolecular forces were at work which could not be calculated by this simple procedure. Hydrogen bonding, probably the strongest type of intermolecular force in a nonelectrolyte, was the clue for making solubility parameter theory work.

No satisfactory method for calculating hydrogen bond interactions was available, so the early applications of this principle depended on the qualitative arrangement of solvents into three hydrogen-bonded classes: alcohols, acids, and amines (strongly hydrogen bonded); esters, ketones, and ethers (moderately hydrogen bonded); and hydrocarbons (poorly hydrogen bonded). In spite of the crudity of this procedure, the prediction of solubility based on solubility parameters within each class is 90% successful.

Because the energy of vaporization of polymers cannot be obtained without decomposition, the solubility parameter is most easily obtained by testing in solvents with different δ values. This procedure led to the second pragmatic extension of Hildebrand's concept—namely, the determination of a solubility parameter range rather than a single value. By this procedure the actual values of ΔH and ΔS do not need to be known. The point at which ΔF becomes positive is determined by experimentally observing insolubility, and the range reported determines $2(\delta_{polymer} - \delta_{solvent})$.

Much attention has been given to quantifying the hydrogen bonding energies. This was reviewed in 1968 (2). The most satisfactory approach is Hansen's (3). He proposed a three-component parameter such that

$$\delta_*^2 = \delta_d^2 + \delta_p^2 + \delta_h^2$$

where δ_d is the dispersion or van der Waal's component of the intermolecular forces, δ_p is the polar component, and δ_h is the hydrogen-bonding component. This approach has a sound theoretical basis, but the actual values of the three components had to be determined more or less empirically.

Recently, Bagley (4, 5) confirmed Hansen's approach by measuring directly the internal pressure of several solvents. The solubility parameters Bagley obtained corresponded closely to the sum of the dispersion and polar forces that Hansen proposed,

$$P_i^{1/2} = (\delta_d^2 + \delta_p^2)^{1/2}$$

The difference between this value and the value obtained from the heat of vaporization corresponds to Hansen's δ_h or the hydrogen-bonding component. These results are most interesting because they confirm the presence of hydrogen bonding in solvents such as acetone where theoretically there should be none. The pragmatism of Hansen's measurements of δ_h and of the qualitative categorization of a group of "moderately hydrogen bonded" solvents seems, therefore, to be justified. While the works of Hansen and Bagley give us a better understanding of the intermolecular forces with which we have to deal, their usefulness for everyday problems is curtailed by the lack of data on dispersion, and polar and hydrogen bonding components of solubility parameters of polymers. Nearly all the published data for polymers is given in solubility parameter ranges in the three qualitative classes of hydrogen-bonded solvents (6).

Quantifying the Hydrogen Bond

The breaking or formation of hydrogen bonds in a solvent mixture greatly changes the character of that mixture. Likewise, the ability to form (or break) hydrogen bonds with the polymer component greatly affects solubility. We have therefore long needed some method of getting at the net results of all of the possible hydrogen bond changes which can occur when a polymer is combined with a mixed solvent.

A practical and valuable approach is suggested by Nelson (7). He used the Pimentel and McClellan (8) classification of solvents based on the way in which they contribute to hydrogen bonds. These are:

(1) Proton donors [such as chloroform and poly(vinyl chloride)].

(2) Proton acceptors [such as ketones, esters, ethers, aromatic hydrocarbons, and poly(vinyl acetate)].

(3) Donor/acceptors (that is, solvents which can act simultaneously as proton donors and acceptors such as alcohols, carboxylic acids, water, and cellulose nitrate).

(4) Non-hydrogen bonding (such as aliphatic hydrocarbons and polyethylene).

The most favorable condition for solubility occurs when a donor compound is mixed with an acceptor compound; for example, a mixture of chloroform and acetone or poly(vinyl chloride) and tetrahydrofuran. Under these conditions, the enthalpy of mixing is negative, thereby ensuring a negative free energy which guarantees miscibility. The most unfavorable condition occurs with donor/acceptor compounds. Here the internal hydrogen bond must be broken before the compound can engage in hydrogen bonding with another molecule. Since this breaking of the hydrogen bond requires energy, the enthalpy is positive which militates against miscibility.

To handle this situation, Nelson proposed a "net hydrogen bond accepting index." Since alcohols (because they are already internally hydrogen bonded) cannot accept any additional hydrogen bonds and, worse yet, since they require energy to break their internal hydrogen bonds, their contribution to hydrogen bond acceptance is negative. Nelson therefore proposed that in the summation of the hydrogen bonding components of a solvent mixture, the value assigned to alcohols should be subtracted rather than added. This algebraic sum he then called his "net hydrogen bond accepting index." Nelson gives several examples which indicate that his technique provides much better predictability of the solvency of solvent mixtures, particularly for poly(vinyl chloride) and its copolymers.

Nelson found that Gordy's (1) data were based on relatively primitive equipment and that he used benzene as a reference-solvent which, of course, is a proton acceptor. This technique mixes deuterated methanol with a given solvent and measures the shift in the infrared spectrum of the O–D bond. Nelson and his co-workers remeasured these shifts for many solvents using carbon tetrachloride as the reference solvent. These are the best data available to date for quantitatively estimated hydrogen bond strengths. They were computerized, together with other solubility data, and used in a program to determine minimum cost of solvent mixtures to meet the Los Angeles County Rule 66 antipollution law.

Kagiya (*10*) also repeated Gordy's experiment for many solvents (using benzene as a reference solvent) and calculated the electron donating power. Interestingly enough, he found some solvents showed a negative shift in the O–D peak, and CCl_4 had the greatest negative shift. The data, therefore, can be corrected to use CCl_4 as the reference solvent for comparison with Nelson's data. Kagiya went one step further and measured the electron accepting power by determining the shift in the carbonyl peak in acetophenone. In this case, the most negative shift was -5.9 cm^{-1} for tetrahydropyran and the greatest shift was 125.2 cm^{-1} for $SbCl_5$. The competitive power to accept or donate electrons can also be measured in mixtures of solvents.

A new set of solubility parameter data has been provided by Hoy (*11*). He gives tables arranged alphabetically, numerically, and by boiling point in which the solubility parameters were calculated by computer from the Antoine equation extrapolated to 25°C. While these values are quite precise, they are sometimes quite different from previously reported values. They should therefore be used with caution in predicting polymer solubility when using published solubility parameter ranges for polymers (*6*). For the experimenter who wishes to set up his own solubility parameter range data, these tables can be quite useful.

Rheineck (*12*) has proposed a modification of Small's Molar Attraction Constants in which he gives corrections for molar volumes. These values are useful for calculating approximate solubility parameters when only the structural formula of a compound is known.

Prigogine Theory

The importance of Prigogine's (*13*) concepts has been discussed by Doolittle (*14*), Patterson (*15, 16*) and Prausnitz (*17*). Prigogine's theory is useful in gaining a better understanding of the entropy changes which occur in solutions. Doolittle points out that the entropy of mixing is not always positive, or at least that the temperature coefficient of entropy change becomes progressively smaller as the temperature is raised. This may cause a system to become immiscible as the critical temperature is approached. Patterson discusses this in some detail, using a lattice model to explain entropy changes. He also discusses the interrelationship of Prigogine concepts and solubility parameter concepts.

In any event, the pragmatism of the solubility parameter concept is seldom affected by the extreme conditions predicted by the Prigogine theory. There are, however, special cases where this concept should be kept in mind—for example, in high pressure plastic molding operations.

Nelson (7) also comments on the effect of hydrogen bonding on entropy of mixing. Both the solubility parameter and the Flory–Huggins theory assume ideal entropy of mixing; that is, there are no intermolecular forces strong enough to interfere with random mixing. This, of course, is not true where hydrogen bonds are formed and gives added force to his proposal of introducing a special treatment in solvent mixtures containing alcohols.

Flory–Huggins Theory

The pertinence of the Flory–Huggins theory (18) has been discussed in detail elsewhere (2). It continues to be used and is valuable in the study of thermodynamics of dilute solutions. It is of relatively little help in solving engineering and formulation problems.

Lewis Acid Theory

One of the newer approaches being studied is the adaption of the Lewis acid-base theory to molecular interactions. In this application hydrogen bonds or polar interactions are not pertinent but just another manifestation of the general concept of electron sharing. The general theory is well presented by Drago and Matwiyoff (19). Lewis acids are compounds which accept electrons, and Lewis bases are compounds which donate electrons. The theory recognizes the fundamental electronic character of "interaction" and "bonding." These are merely aspects of electron sharing where "bonding" denotes a large energy of electron attraction and "interaction" (such as the molecular interactions existing in mixtures of solvents and other materials) denotes lower energies of electron sharing.

Drago (20, 21) extended this theory to consider that electron sharing is divided into two components: an electrostatic and a covalent force. He and his co-workers have measured the electrostatic E and covalent C components of compounds from which the bonding or interaction energy (ΔH) can be calculated by:

$$\Delta H = E_A E_B + C_A C_B$$

where subscripts A and B refer to two different components being mixed. The advantage of this treatment is that it should handle not only interactions which we normally call hydrogen bonding or polar, but also ionic interactions, aqueous solutions, and metal–organic interactions. It would be desirable to have one comprehensive theory encompassing metals, electrolytes, and nonelectrolytes.

Drago has suggested that this could be applied to polymer solutions by hypothesizing a "reaction" such as

$$P + S = SP + S$$

where P equals polymer, S equals solvent, and SP equals "solvated polymer." The enthalpy of this reaction could theoretically be calculated from the E and C values of P and S.

Unfortunately, no values for polymers have been measured in this system. There is no provision for handling dispersion forces which are of great importance in many practical systems. Nevertheless, as this theory is developed it should be of value in handling aqueous solutions, perhaps latex stabilization, film formation, adhesion, or pigment-vehicle interactions.

A great deal of literature exists where workers have studied acid-base characteristics or electron or proton donor characteristics of solvents as they relate to reaction rates. These have been reviewed by Dack (22) and Reichardt (23). Although these studies are interrelated, they have resulted in many different empirical solvent classification schemes leading to ϵ-values, X values, Y values, Z values, E_T values, R and S values, Δv_D and Δv_A, etc. These papers may be a source of data, but they are of little value in understanding solvent action in coatings and plastics. One interesting proposal, however, was to use "diphenyl betain" as a polarity indicator (24); this is a dye which changes color when dissolved in different solvents, varying from red (in methanol) to green (in acetone).

Molecular Clustering Theories

A completely fresh and different approach to polymer solutions is being studied in at least two places under the auspices of the Paint Research Institute. Some years ago, Prausnitz (25) reported on polymer segment-interaction theory to predict phase separation. This was based on a lead by Wilson (24) who proposed a local volume fraction concept or domain theory for explaining deviations from ideality. More recently, Prausnitz (27) reported on further work involving anomalous viscosities in high solid polymer solutions. High viscosities may occur when solvent mixtures are used which have values near the extreme ends of the solubility parameter ranges for polymers. This is caused by aggregation or clustering of polymer molecules before phase separation. It is probable also that these clusters are not completely broken down to individual molecules even with "good" solvents. There is always some degree of entanglement or aggregation even in solvents with solubility parameters in the middle of the polymer ranges. This phenomenon manifests itself in alkyd resin solutions where average molecular weights of 2000–3000 may produce viscosity average molecular weights of 100,000–300,000. This phenomenon is ill understood, and an extension of the local volume fraction concept is badly needed.

A related but again different approach has been started by Rudin (28). His concept is basically one of colloid chemistry in considering the conditions which would lead to coagulation. Assuming that one has a dispersion of polymer particles, not necessarily monomolecular, the speed with which these particles will agglomerate to particle sizes in the visible range depends on their diffusion rate. Rudin has been able to relate diffusion constants to the K and α constants in the Mark–Howink equation

$$[\eta] = KM_V{}^\alpha$$

For a given polymer solvent system, many values of K and α are reported. If Rudin is successful in extending this approach, it may be possible to predict solubility from data already known.

Technology

Numerous applications of solubility parameter theory to practical problems have been reported (2, 29). Recently, Hansen and Beerbower (30) have presented a concise summary of the present state of theory and practice, including thixotropy, dispersion coatings, emulsions, and latexes. They also discuss such non-coating applications as swelling of rubber gaskets, liquid-liquid extractions, interface studies, solubility of gases, high pressure gas chromatography, and the effect of temperature and pressure on solubility parameter. This reference also has the most complete and up-to-date values of Hansen's three components of solubility parameter.

An active field is interfacial free energy. Patton discussed the importance of surface tension on flow, adhesion, wetting, film formation, and dispersion as related to solvents and coatings in his Mattiello Lecture (31). Hahn (32) has been able to relate many coating defects such as cratering, wrinkling, flooding and floating, and picture framing to changes in interfacial energy when solvents evaporate from films. Hansen (33) pointed out the relationship of solubility parameter to wetting of surfaces while Wu (*34*) showed how to calculate Zisman's critical surface tension from Small's molar attraction constants. Lee (*35*) confirmed Hildebrand's relationship of solubility parameter with surface tension of solvents. He determined values of K and α in the equation

$$\delta = K \left(\frac{\gamma}{V_M^{1/3}} \right)^\alpha$$

for 14 types of solvents and found K to vary from 3.42 to 5.96 and α from 0.25 to 0.58; the variation was greatest in oxygen-containing solvents. Lee also tabulates actual surface tension values for many solvents.

Beerbower (*36*) has correlated solubility parameter with emulsifier selection with some success. Following Winsor (*37*), he calculates a ratio of the lyophobic to hydrophilic portions of emulsifiers using Hansen's three-component solubility parameter values. In the one test reported, there seems to be excellent correlation of the optimum ratio with stability of the emulsion.

Another active field is that of pigment dispersion. Some of the problems connected with theory and practice have previously been discussed (*2, 38*). Continuing the work of Hansen (*30*) and Sorensen (*39*), Eissler (*40*) studied the sedimentation of a zinc oxide pigment, both untreated and treated with a proprietary octyl phosphate. There was an optimum δ_H for the untreated pigment (δ_d and δ_p had little effect). The polar factor correlated with the treated pigment, the sedimentation volume increasing with increasing δ_p. The Technical Committee of the New York Society for Paint Technology is also studying this relationship with the hope of establishing a wetting parameter for pigments.

The accomplishments in the past decade have been considerable. Much remains to be done. One wonders, however, whether ecological factors will not so promote the use of powders, films, and aqueous and high solids coatings that the incentive for further study of solvents will be greatly curtailed.

Literature Cited

1. Hildebrand, J. H., Scott, R. L., "The Solubility of Nonelectrolytes," Reinhold, New York, 1950.
2. Burrell, H., *J. Paint Technol.* (1968) **40**, 197.
3. Hansen, C., *J. Paint Technol.* (1967) **39**, 104, 511.
4. Bagley, E. B., *et al., J. Paint Technol.* (1969) **41**, 495.
5. Bagley, E. B., *et al., J. Paint Technol.* (1971) **43**, 35.
6. Burrell, H., "Polymer Handbook," Sec. IV, p. 341, Interscience, New York, 1966.
7. Nelson, R. C., *et al., J. Paint Technol.* (1970) **42**, 636, 644.
8. Pimentel, G. C., McClellan, A. L., "The Hydrogen Bond," Freeman, San Francisco, 1960.

9. Gordy, W., *J. Chem. Phys.* (1939) **7**, 93.
10. Kagiya, T., et al., *Bull. Chem. Soc. Jap.* (1967) **41**, 767–782.
11. Hoy, K. L., *J. Paint Technol.* (1970) **42**, 76.
12. Rheineck, A. E., Lin, K. F., *J. Paint Technol.* (1968) **40**, 611.
13. Prigogine, I., "The Molecular Theory of Solutions," Interscience, New York, 1957.
14. Doolittle, A. K., *J. Paint Technol.* (1969) **41**, 483.
15. Patterson, D. D., *J. Paint Technol.* (1969) **41**, 489.
16. Patterson, D. D., *Rubber Chem. Technol.* (1967) **40**, 1.
17. Prausnitz, J. M., *Chem. Eng. Sci.* (1965) **20**, 703.
18. Flory, P. J., "Principles of Polymer Chemistry," p. 508, Cornell University Press, Ithaca, 1953.
19. Drago, R. S., Matwiyoff, N. A., "Acids and Bases," a volume in the "Topics in Modern Chemistry Series," Heath, Lexington, Mass., 1968.
20. Drago, R. S., "Symposium on Paint Phenomena Influenced by Acid–Base Interactions," 1970.
21. Epley, T. T., Drago, R. S., *J. Paint Technol.* (1969) **41**, 500.
22. Dack, M. R., *J. Chem. Technol.* (1971) **1**, 108.
23. Reichardt, C., *Angew. Chem., Int. Ed. Engl.* (1965) **4**, 29.
24. Dimroth, K., et al., *Justus Liebigs Ann. Chem.* **661**, 1–37.
25. Heil, J. F., Prausnitz, J. M., *AIChE J.* (1966) **12**, 678.
26. Wilson, G. M., *J. Amer. Chem. Soc.* (1964) **86**, 127.
27. Prausnitz, J. M., AIChE Meeting, Cleveland, May 4, 1969, Paper 3C.
28. Rudin, A., *J. Paint Technol.* (1972) **44**, 41–60.
29. Burrell, H., *Off. Dig. Fed. Soc. Paint Technol.* (1955) **27**, 726.
30. Hansen, C., Beerbower, A., "Solubility Parameters," in "Encyclopedia of Chemical Technology," Suppl. Vol., Interscience, New York, 1971.
31. Patton, T. C., *J. Paint Technol.* (1970) **42**, 666.
32. Hahn, F., *J. Paint Technol.* (1971) **43**, 58–67.
33. Hansen, C., *J. Paint Technol.* (1970) **42**, 660.
34. Wu, S., *J. Phys. Chem.* (1968) **72**, 3332.
35. Lee, L-H., *J. Paint Technol.* (1970) **42**, 365.
36. Beerbower, A., Nixon, J., *Amer. Chem. Soc., Div. Petrol. Chem., Prepr.* (1969) **14**, 62.
37. Winsor, P., "Solvent Properties of Amphiphilic Compounds," Butterworth, London, 1954.
38. Burrell, H., *J. Paint Technol.* (1970) **42**, 2.
39. Sorensen, P., *J. Oil Colour Chem. Ass.* (1967) **50**, 226.
40. Eissler, R. L. et al., *J. Paint Technol.* (1970) **42**, 483.

RECEIVED March 14, 1972.

2

Predicted Compositions during Mixed Solvent Evaporation from Resin Solutions Using the Analytical Solutions of Groups Method

E. L. DERR and C. H. DEAL

Shell Development Co., 3737 Bellaire Boulevard, Houston, Texas 77001

> *Activity isotherms for solvent-resin solutions are correlated, and compositional trends during evaporation of mixed solvents are predicted in terms of assigned size and structural groups and group-pair parameters using the ASOG method. The study relates to alkyd resins of varying methylene/aromatic ratios and solvents including paraffin, aromatic, chloride, ketone, ester, and alcohol types. The thermodynamic nature of ASOG allows the prediction of both resin and solvent activities and thereby furnishes a basis for miscibility estimates. Parameters generated can be used in other homologous systems involving the groups studied here (the same types of resins and solvents).*

Film drying takes place in two stages (1). In the first stage, where 85–95% of the solvent evaporates, the rate is controlled by a vapor barrier; vapor composition at this point approaches that expected in equilibrium vaporization. Changes in composition of the residual liquid can therefore be calculated in the same way that changes in the liquid composition from a flash distillation can be calculated. This determination is important to the formulator since the residual solvent composition partially determines the nature of the set surface; gloss, for instance, hinges on continued miscibility and attendant leveling action.

Therefore, equilibrium solution behavior during drying is important. In drying solvents from alkyd resins, there is an almost one-to-one correspondence between equilibrium effects and observed results. However, even in cases where equilibrium effects are only a part of the picture, knowledge of solution behavior contributions contribute to an understanding and a more accurate appraisal of the controlling diffusional effects such as those in the liquid which dominate the second stage of drying. Solution behavior is defined in terms of component activity coefficients; these coefficients define miscibility regions and are used in calculating resin solution viscosity.

Several years ago, an extended experimental program was carried out by G. M. Sletmoe (Shell Development Co.) geared towards helping the coatings formulator and pressmen to utilize solvent blends effectively. These studies involved determination of solvent evaporation rates for both neat solvent blends and alkyd resin solutions. Sletmoe also determined isotherms showing solvent activity in resin solutions through isopiestic measurements to relate evaporation rates to equilibrium solvent partial pressures. Some of these data have been published (2, 3, 4) with guideline observations and rules for their application.

The present paper correlates all of Sletmoe's alkyd resin solution data into a uniform thermodynamic package using the analytical solution of groups (ASOG) method (5, 6).

The Hildebrand solubility parameter method, modified for polarity and hydrogen bonding effects, has been used to correlate and define surface coating solvent–polymer miscibility and viscosity properties (7, 8). The volatilities of neat solvent blends have been correlated according to the van Laar relationship (9). A satisfactory thermodynamic treatment should represent both solvent volatility and miscibility regions. Some success was achieved in this regard using the Flory-Huggins expression to describe the athermal contributions of such systems and an expanded Wilson expression describing the energy, heat or χ term (10). The χ constant is found to vary with composition in the classical use of the Flory-Huggins expression, and Ref. 10 accounts for this as a function of interactions between monomer segments of the polymer and the monomer segments of solvent molecules.

The ASOG method predicts χ and its concentration dependency on the basis of interactions between the segments or groups of the polymer and the segments or groups of the solvent molecules. Rough miscibility mapping requires a less precise description of the polymer–solvent system than does the calculation of solvent partial pressures. Thus, a pattern for the successful calculation of partial pressures can describe reasonable solubility limits; for instance, when calculated activities exceed values of one, a demixing situation can be expected.

Relationships and ASOG Matrix Development

The present calculations of activities and compositions have been made using Equations 1–7; group counts or assignments are given in Table I and group-pair parameters are given in Table II.

$$P_i = a_i P_i^\circ = x_i \gamma_i P_i^\circ = y_i P \qquad (1)$$

$$\log \gamma_i = \log \gamma_i^{FH} + \log \gamma_i^{G} \qquad (2)$$

$$\log \gamma_i^{FH} = \log R_i + 0.434(1 - R_i) \qquad (3)$$

$$R_i = \nu_{FHi} / \sum_i x_i \nu_{FHi} \qquad (4)$$

$$\log \gamma_i^{G} = \sum_k \nu_{ki} \log \Gamma_k - \sum_k \nu_{ki} \log \Gamma_k^{i} \qquad (5)$$

$$\log \Gamma_k = -\log \sum_l X_l a_{kl} + 0.434 \left[1 - \sum_l \frac{X_l a_{lk}}{\sum_m X_m a_{lm}} \right] \qquad (6)$$

$$X_l = \frac{\sum_i x_i \nu_{li}}{\sum_i x_i \sum_l \nu_{li}} \qquad (7)$$

Equations 2–7 break down the ASOG correlation as described in Ref. 5. The activity coefficient γ_i is the product of two activity coefficients, an athermal size ratio γ_i^{FH} and an interaction γ_i^{G} (Equation 2). The athermal γ_i^{FH} is calculated with a Flory-Huggins type expression (Equation 3) where R_i is the ratio of molecule i size to the average size of the molecules in solution. These sizes are assigned as ν_i^{FH} and are listed under FH in Table I. The interaction γ_i^{G} is calculated in terms of group activity coefficients Γ_k and the number of such groups in molecule i denoted as ν_{ki}, as in Equation 5. In Equation 5, Γ_k is the group k activity coefficient in the solution of interest and Γ_k^{i} is the group k activity coefficient in the environment of pure compound i. The group activity coefficients Γ_k are calculated in terms of parameters for group pairs a_{lk} and group concentrations X_k using the Wilson expression as in Equation 6. Group concentrations are a function of mole fraction concentrations, x_i, and the number of groups associated with the various molecules ν_{ki}, as in Equation 7. Interaction groups ν_{ki} are assigned along with the previously mentioned size assignments; the assignments used in the present calculations are given under the respective groups in Table I. The parameters a_{lk} for the present calculations are given in Table II. Group counts or assignments generally follow the nature of the molecule but not rigorously so; however, assignments and interaction parameters are rigorously interrelated.

Table I. Properties and Group

Substance	MW	ρ	$P°$, mm	CH_2
Resin				
Beckosol 34	2100	1.12	0	64.0
Resins for Comparison				
Beckosol 34	2100	1.12	0	64.0
Beckosol 31	(2100)	1.07	0	71.0
Long	2175	1.048	0	82.0
Medium	2050	1.120	0	62.0
Short	2025	1.181	0	51.0
Monomers				
DOP	390.54	0.986	0	16.0
SAM[a]	351.0	0.78(?)	0	9.9
Solvents				
Nonane	128.0	0.72	4.6	13.0
Heptane	100.2	0.68	35.7	11.0
Toluene	92.13	0.868	28.4	1.0
o-Xylene	106.2	0.876	6.7	2.0
Cumene	120.0	0.858	4.4	3.0
IPAC	102.1	0.877	47.3	4.0
MAAC	144.2	0.86	5.2	7.0
MIPK	86.13	0.803	39.6	4.0
MIAK	114.0	0.81	2.8	6.0
IPA	60.08	0.786	31.8	3.0
BuOH	74.0	0.81	6.44	4.0
Chlorobenzene	112.56	1.106	11.75	—

[a] 50–50 wt % dioctylphthalate (bis-2-ethylhexyl) and butylbenzylphthalate.

Table II. Group-pair Parameters, and

Parameters

	CH_2	CA	OH	$MCOO$
CH_2	1.0000000	1.0838000	1.2790000	1.8490000
CA	0.8505000	1.0000000	1.1588000	0.4701000
OH	0.0049300	0.0213500	1.0000000	0.4801000
$MCOO$	0.1453300	0.8496700	0.2963000	1.0000000
KCO	0.0250000	0.6075000	0.4719000	1.0386000
$DCOO$	0.0473000	0.0364800	0.0000400	0.0017120
Cl	0.1040000	1.0000000	0.0794000	2.0186000

Limiting Group

	CH_2	CA	OH
CH_2	1.0	1.07	2.1
CA	1.08	1.0	2.3
OH	153.4	40.0	1.0
$MCOO$	2.9	2.0	5.7
KCO	12.5	2.3	2.8
$DCOO$	44.7	14.2	36950.0
Cl	2.2	1.0	25.4

Counts Used in Correlations

CA	OH	MCOO	KCO	DCOO	Cl	FH
48.0	3.0	4.0	0.0	8.0	0	100.0
45.2	1.8	4.5	0.0	9.1	—	102.5
39.2	2.1	4.28	0.0	8.57	0	100.0
37.0	1.6	4.0	0.0	8.0	0	113.3
48.0	2.9	4.5	0.0	9.07	0	100.0
46.2	2.5	5.1	0.0	10.3	0	94.3
6.0	—	—	—	2.0	—	28.0
9.33	—	—	—	2.0	—	25.23
—	—	—	—	—	—	9, 5.6
—	—	—	—	—	—	7.0
6.0	—	—	—	—	—	7.0
6.0	—	—	—	—	—	8.0
6.0	—	—	—	—	—	9.0
—	—	1.0	—	—	—	7.0
—	—	1.0	—	—	—	10.0
—	—	—	1.0	—	—	5.0
—	—	—	1.0	—	—	8.0
—	1.0	—	—	—	—	4.0
—	1.0	—	—	—	—	5.0
6.0	—	—	—	—	1.	7.0

Associated Limiting Group Activity Coefficients

Parameters

KCO	DCOO	Cl
2.1632999	0.2514000	2.4500000
0.6788000	1.6573000	1.0000000
0.7333000	0.6093000	0.3000000
0.9176000	1.5780000	0.0450000
1.0000000	0.7171000	1.0000000
0.1585000	1.0000000	0.0743000
1.0000000	0.0209000	1.0000000

Activity Coefficients

MCOO	KCO	DCOO	Cl
1.27	1.23	10.3	1.0
2.47	2.18	1.58	1.0
4.2	2.31	4.46	8.4
1.0	1.05	1.72	8.0
1.05	1.0	3.23	1.0
327.0	8.4	1.0	35.8
1.3	1.0	120.7	1.0

The ASOG method would be impractical without electronic computers and appropriately designed programs. With these facilities available, the use of ASOG involves two essential steps:

(1) Breakdown of molecules of interest into reasonable structural group and size factors.

(2) Establishing parameters which represent behavior of these groups from appropriate activity coefficient data. These data-based systems may or may not contain molecules of interest but must contain groups or segments of interest.

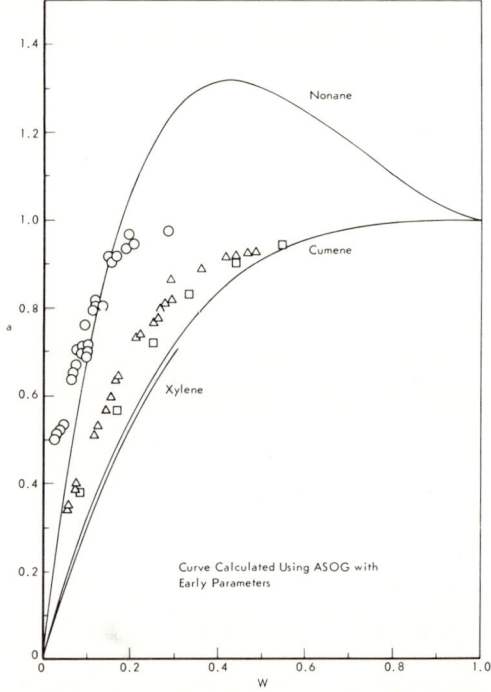

Figure 1. Activities of nonane, xylene, and cumene from Beckosol 34 resin

For the present calculations, the compounds and assigned group breakdowns are listed in Table I. We are concerned with mixtures having ketone KCO, ester MCOO, diester DCOO, alcohol OH, and chloride groups. In each of the solvents we count saturated carbons as being identical to CH_2 and aromatic carbons identical to CA. In paraffins we count each end group equal to $3CH_2$ to account for end effects. Otherwise, in this calculation each designated group is assigned effective counts of one. (Later work has shown the advisability of counting according to number of non-hydrogen atoms making up the group;

thus, the MCOO could better be given a value of three in the methyl amyl acetate solvent, but here interaction parameters are based on the one count, and this count must be adhered to. The ASOG method must be recognized as a developing procedure; present assignments and associated parameters are in no way final.)

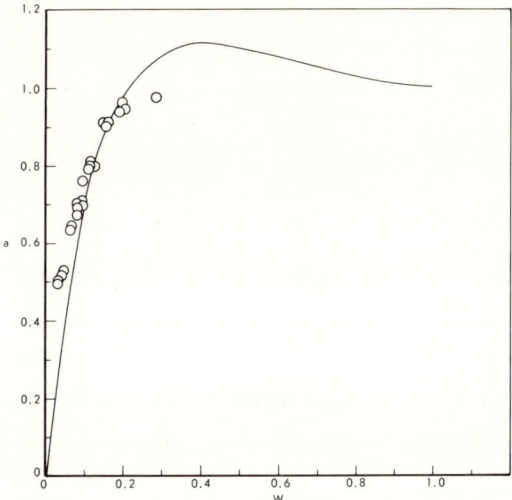

Figure 2. *Activity of nonane in Beckosol 34 resin*

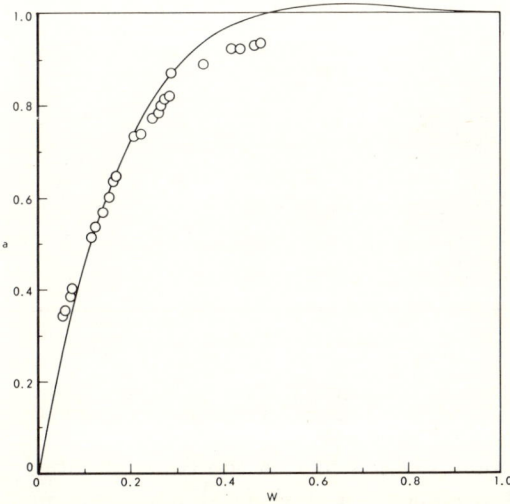

Figure 3. *Activity of cumene in Beckosol 34 resin*

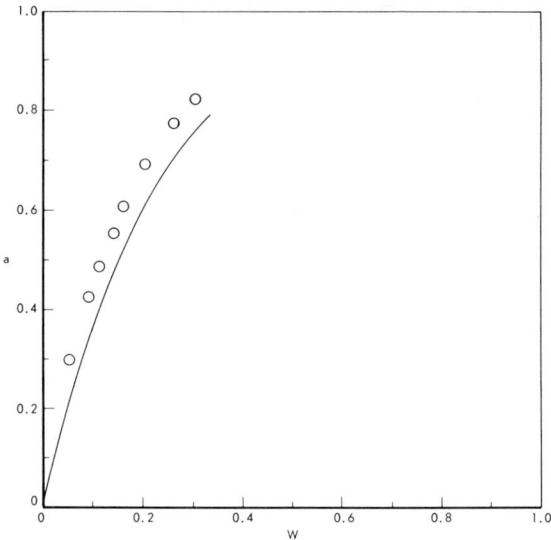

Figure 4.   Activity of MIPK in Beckosol 34 resin

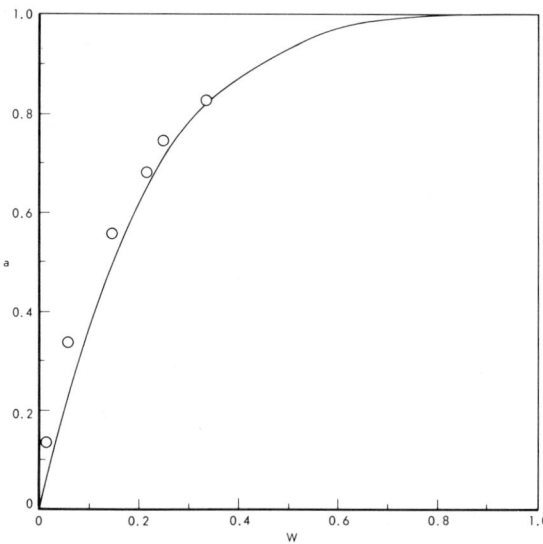

Figure 5.   Activity of IPAC in Beckosol 34 resin

The group count assignments for the Beckosol alkyd resins are based on estimated stoichiometry and their ebullioscopic molecular weights. The resin *FH* counts are calculated using a value of 100 for Beckosol 34 and the ratio of the various mole weights.

The group pair parameters are summarized in Table II. Parameters for CH_2–CA, CH_2–OH, CH_2–MCOO, CH_2–KCO, and CH_2–Cl were computer generated by satisfying activity coefficient data (*11*) for heptane–benzene, heptane–ethanol, hexane–ethyl acetate, heptane–acetone, and heptane–chlorobenzene systems, respectively. In generating the CH_2/Cl parameters, previously determined CH_2/CA values were used. In order that no more than a single pair of parameters be sought in a given determination, it is important that proper sequence be maintained.

The master activity curves determined by Sletmoe for solvents in dioctylphthalate and in alkyd monomer simulator were used in the present calculation to determine parameters for DCOO or the ester groups attached to the benzene rings of the alkyd backbone.

A sublisting in Table II shows calculated group limiting activity coefficients based on the developed matrix to help gain a physical feel for the parameter values. Values for the DCOO group are generally higher than seems intuitively reasonable, especially the value reflecting the DCOO/OH interaction. This can suggest weakness in the master isotherm data, supportive parameters, or the model itself.

Resin Isotherms

An early success of ASOG in representing solvent–resin behavior is shown in Figure 1 by the prediction of the activity of nonane in Beckosol 34. At that time, we had parameters for methylene CH_2, olefin CH,

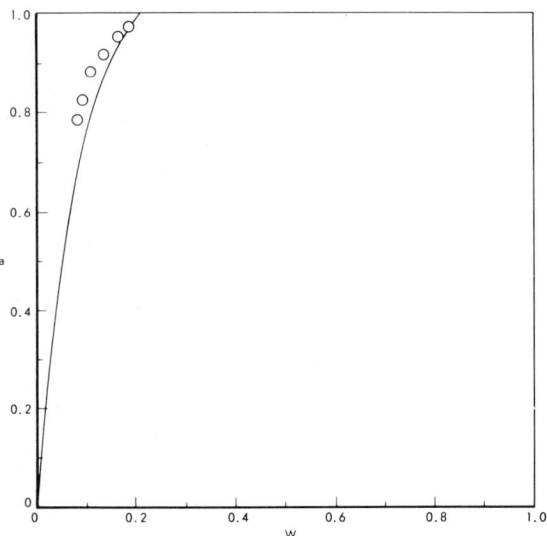

Figure 6. Activity of IPA in Beckosol 34 resin

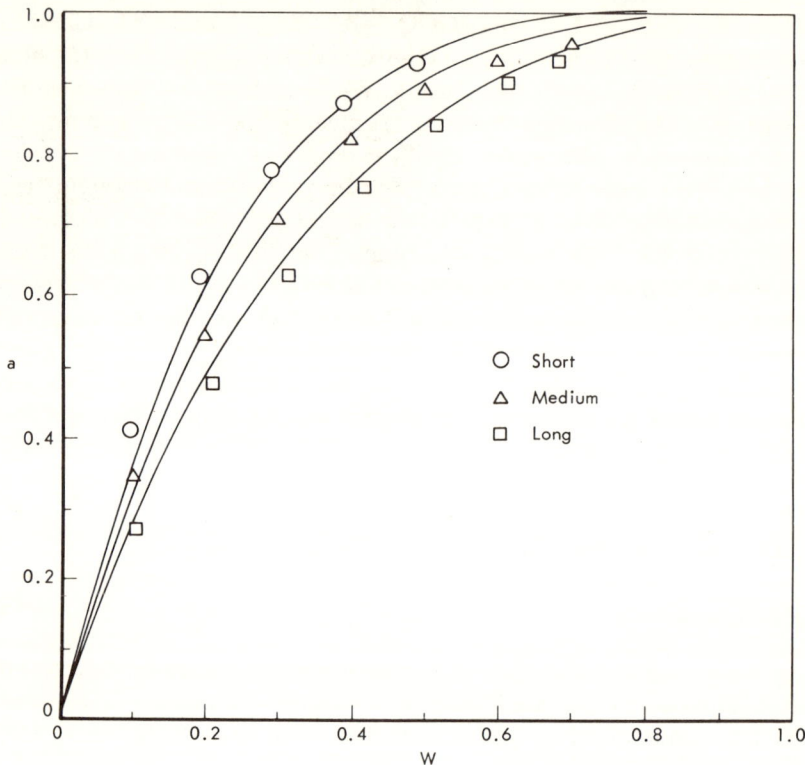

Figure 7. Activity of chlorobenzene in various alkyd resins

hydroxyl OH, ether O, and ketone C=O. To work within this limited array of parameters, we equated olefin CH and CA groups and artificially constructed the ester COO group by combining the ether O and ketone C=O groups. Tests with cumene solvent were somewhat off as shown, but tests with oxygenated solvents clearly indicated the desirability for a modified matrix. In this preliminary calculation, a Flory-Huggins count of 150 was used to correspond roughly to the number of nonproton atoms based on the ebullioscopic molecular weight of the resin.

On generating CA, MCOO, and DCOO parameters, it was observed that solvent activity representation was generally improved, but except for nonane, predicted solvent activities were consistently low. This indicated an over influence of the Flory-Huggins term, and effective counts of 100 and 5.6 were used for the resin and nonane components in the final correlation. The reduction of the entropy contribution for the resin can be justified on the basis that orientative effects are operating; the behavior of nonane is somewhat anomalous.

Figures 2, 3, 4, 5, and 6 are the ASOG depictions of the isotherms for nonane, cumene, methyl isopropyl ketone, isopropyl acetate, and isopropyl alcohol in Beckosol 34. These results are based entirely on parameters determined from data for small molecules except for the above resin, nonane size adjustments. Better individual fits could have been made, but at a loss in the general predictive character of the method. General observations might include a tendency towards predicting low activities at high resin levels and somewhat high activities at the 50% level for nonane, slightly high values at high cumene levels, generally low ketone activities, satisfactory ester predictions, and slightly low alcohol activities again at high resin levels.

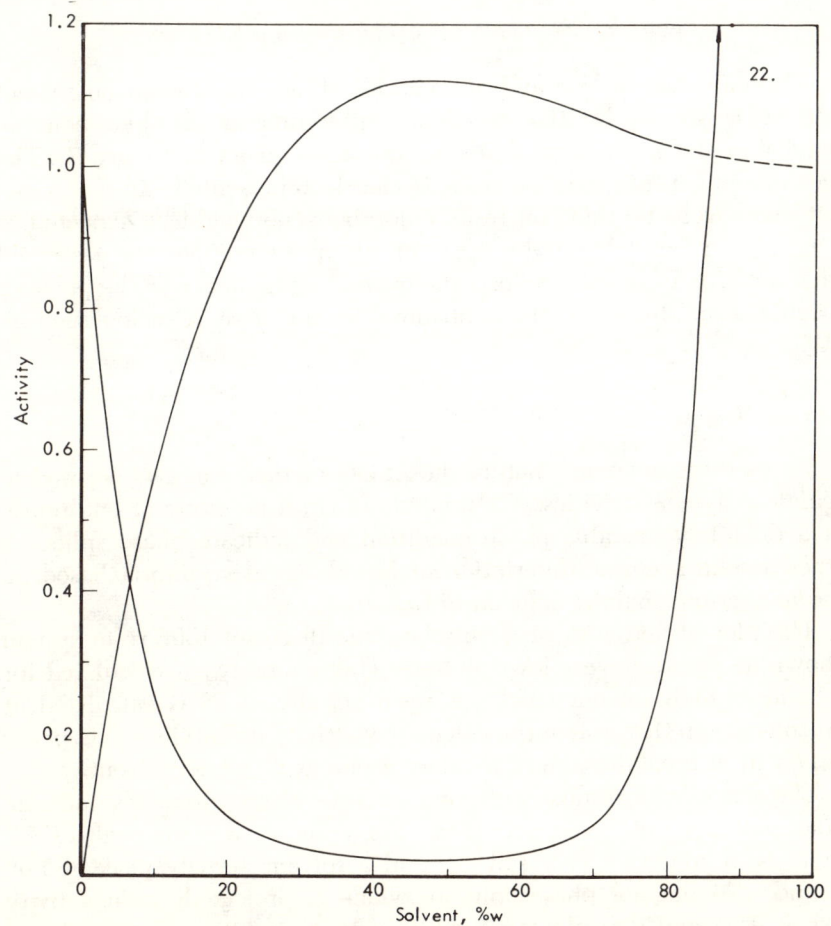

Figure 8. Activity isotherm: butyl benzene–Beckosol 1307

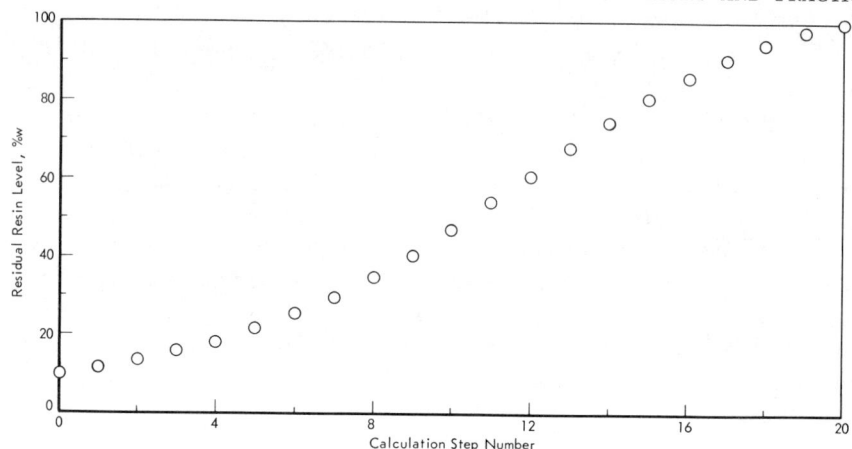

Figure 9. Step pattern in vaporization calculation

The influence of size and aromaticity of the alkyd resin on solvent behavior is shown by the excellent representation of chlorobenzene behavior in short, medium, and long oil resins given in Figure 7. The lower activity in the long oil resin is clearly represented. In the somewhat more aromatic short oil resin, chlorobenzene would be expected to have a lower activity on the basis of group-interaction; the observed higher activity apparently reflects the overriding influence of the polymer molecule size. Both factors contribute to the activity calculation in ASOG.

Resin Activity

By its thermodynamic nature the ASOG method can also be used to calculate polymer activities. Calculated activities in excess of one represent a definitely unstable phase condition and indicate phase splits. A method for an accurate description of liquid–liquid equilibria based on rigorous thermodynamics is given in Ref. 10.

The plot of activities of a butylbenzene-Beckosol 1307 resin system is shown in Figure 8; very low solubility (high activity) is calculated for the resin at high solvent levels. A resin activity of 22 is calculated at 90% solvent. In this system the calculated activity of butylbenzene peaks at 1.17 with a breakthrough of a value of one at 27 wt % solvent.

The calculated results with smaller side chain aromatics indicate similar patterns but lower activities. For cumene, a resin peak of 2.7 activity is obtained at 90% solvent, and a solvent activity peak of 1.08 is found. No definite phase split in xylene is projected; resin activity peaks at 0.23, and the calculated xylene activity is 1.05 and quite flat at 55 wt %.

The calculation of resin and solvent activities has had a recent application in the correlation and prediction of mixture viscosities (*12*) which should prove valuable to the solvent formulator.

Evaporation Tests

Sletmoe's evaporation studies with mixtures of solvent blends and alkyd resins offered only guideline comparisons between expected equilibrium vaporization and his experimental results. His data provide an excellent basis for evaluating the use of ASOG in quantitatively predicting compositional fates during alkyd wet-drying.

These studies were conducted with binary solvent blends of nonane–cumene, nonane–methyl isoamyl ketone, nonane–methyl amyl acetate, and nonane–butanol, starting with 40 wt % solvent-60 wt % resin. Runs with starting solvent blends of 10, 25, 50, 75, and 90 wt % nonane were carried out to generally less than 5 wt % residual solvent. Compositions

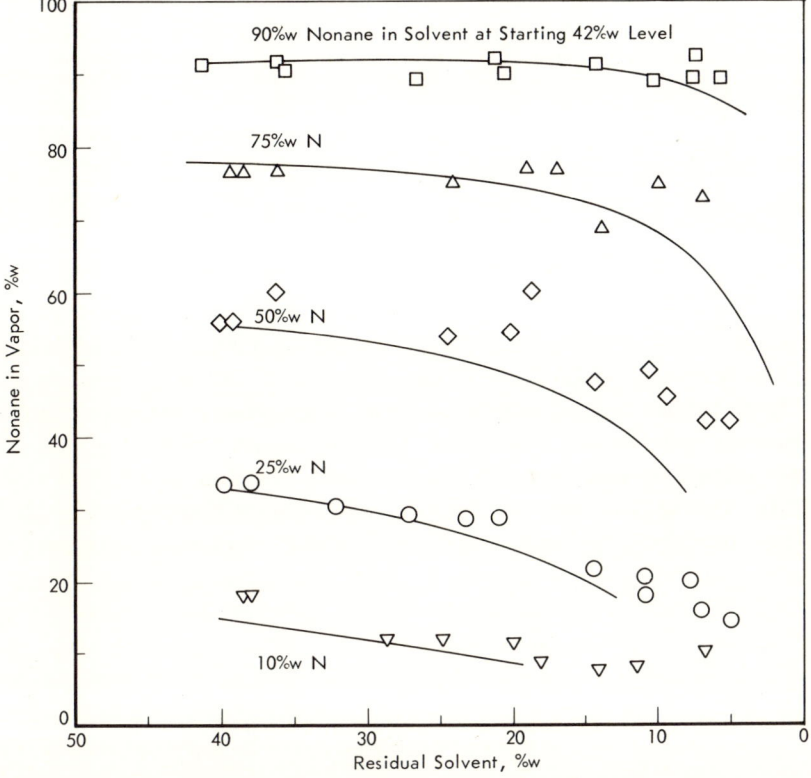

Figure 10. Vaporization of nonane–cumene blends from Beckosol 34 resin

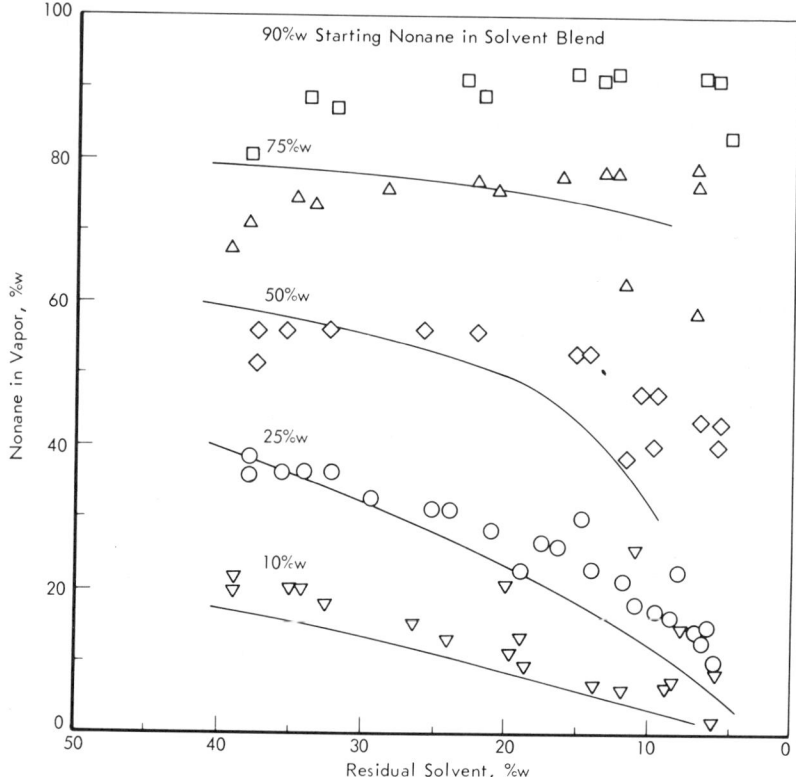

Figure 11. Vaporization of nonane–MIAK blend from Beckosol 34 resin

of the vapor were followed by GLC analysis in all tests, and spot liquid solvent blend compositions were measured in the 50 wt % nonane tests.

Our predicted compositions were determined using a variably stepped material balance program. Partial pressures are calculated from in-put liquid compositions and vapor pressures and ASOG calculated activity coefficients. This results in a new composition according to the routines described by Equations 8 and 9, and the process is repeated.

$$y_i = \frac{P_i}{\Sigma\, P_i} \quad (8)$$

$$x_i^{new} = \frac{\left[x_i - y_i \frac{(YF)\,(NREP)}{(IREP)\,+\,(NREP)} \right]}{\sum\limits_i \left[x_i - y_i \frac{(YF)\,(NREP)}{(IREP)\,+\,(NREP)} \right]} \quad (9)$$

Each step causes the removal of a certain amount of this vapor from the liquid depending on the programmed fraction, YF, the step number $NREP$ and the total number of steps $IREP$ planned for the complete vaporization. Mostly 20 iteration steps $IREP$ and a fraction depletion YF of 0.13 were used in these calculations. More steps of smaller fractions can, of course, be used. The above function is designed to give sensitivity to initial and final changes during vaporization from polymer when viewed on a weight basis. This is shown by the diagram in Figure 9.

Calculated and experimental trends in the vapor composition from nonane–cumene–Beckosol 34 are shown in Figure 10. A tendency to initial higher nonane concentrations in the vapor reflecting the higher volatility of this component is observed. The trends in composition with evaporation are predicted by calculation: a more rapid drop in nonane content occurs during evaporation as the initial nonane percentage in the solvent decreases.

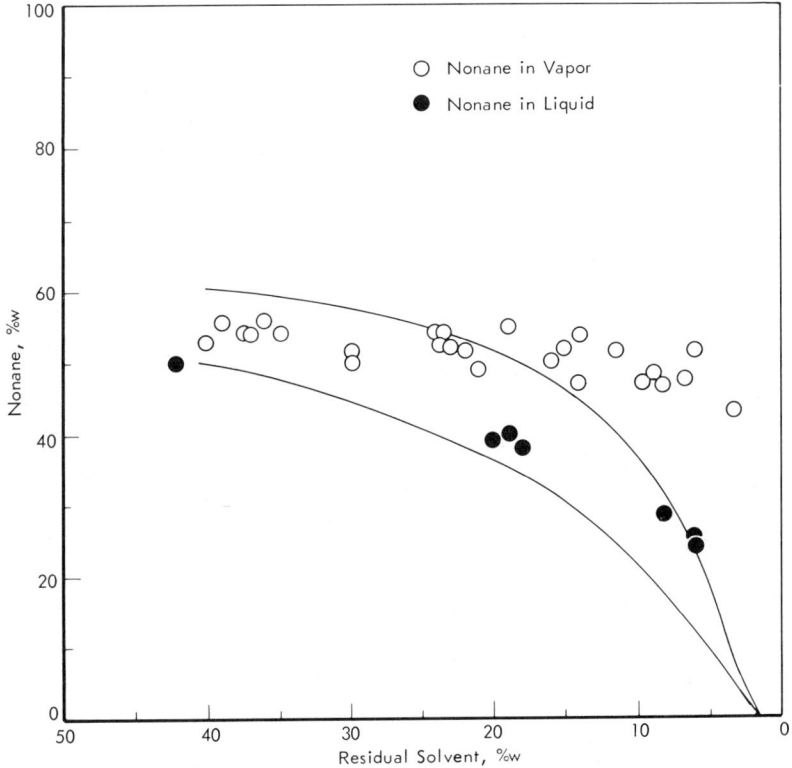

Figure 12. *Vaporization of nonane–MIAK blend from Beckosol 34 resin*

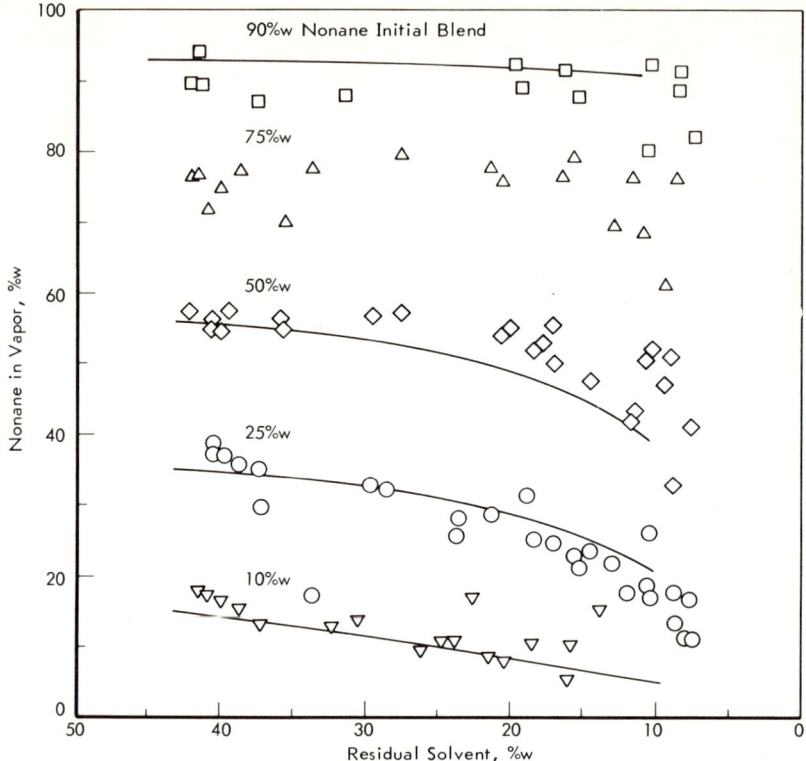

Figure 13. Vaporization of nonane–MAAC blend from Beckosol 34 resin

Vapor composition trends are again anticipated by calculation for nonane–methyl isoamyl ketone blends in Figure 11. Predicted compositions are less exact for the 50 wt % blends than for the blends containing both more and less nonane. This is typified by the plot in Figure 12 where experimental and calculated vapor and liquid compositions are each compared. Deviations observed are in line with deviations in isotherm representation: high nonane, low ketone activities at high solvent levels, and low nonane activity at low solvent levels. Data for nonane–methyl amyl acetate blends compare closely with calculated values in Figures 13 and 14.

An example of vapor composition prediction is shown for the nonane–butanol case in Figures 15 and 16. At starting levels of solvent having greater than 50 wt % nonane, the nonane in the vapor increases as the vaporization progresses; where the initial concentration is less than 50 wt % the nonane content decreases with drying. This reflects the activity coefficient changes: at high levels of nonane, butanol undoubtedly has a substantial γ, and at initial high butanol levels nonane has initial high γ's and volatiles. A curious trend in nonane concentration is observed for the 50% initial blend (Figure 16): perfect balance is noted during the 40–24 wt % solvent level, but then a rather marked increase in nonane level occurs, reaching a maximum at 10 wt % residual solvent. This trend is not anticipated by the equilibrium considerations and must be ascribed to diffusional effects or possibly unpredicted phase splits. Even so, the calculated nonane content in the 7% residual solvent is 40% compared with an experimental value of 30%.

The compositional projections for methyl isoamyl ketone, methyl amyl acetate, and butanol are gratifying in that these specific compounds were never involved in the base data upon which the ASOG matrix is built. This specifically illustrates simultaneous composition and structure projection.

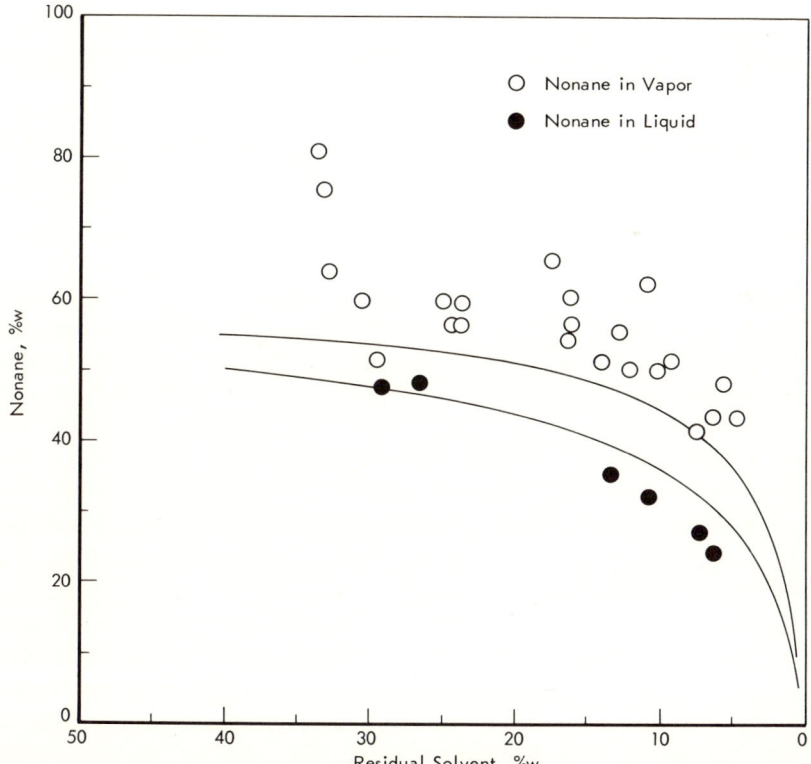

Figure 14. Vaporization of nonane–MAAC blend from Beckosol 34 resin

Conclusions

The ASOG method, developed to correlate activity coefficients in multi-stage, multi-component separation processes, furnishes reasonable vapor and liquid compositions for the wet stage of mixed solvent evaporation from alkyd resins solutions. The versatility of the method may prove advantageous in predicting the properties for solvent–polymer mixtures involving strong specific interactions. In addition to solvent volatility, ASOG provides the basis for calculating polymer activities. Polymer activity can indicate adequate or inadequate solvency and represent a necessary parameter in the correlation of viscosities of solvent–polymer solutions. Established group-associated parameters allow not only the prediction of specific mixture behavior but most homologous variations of such mixtures as well.

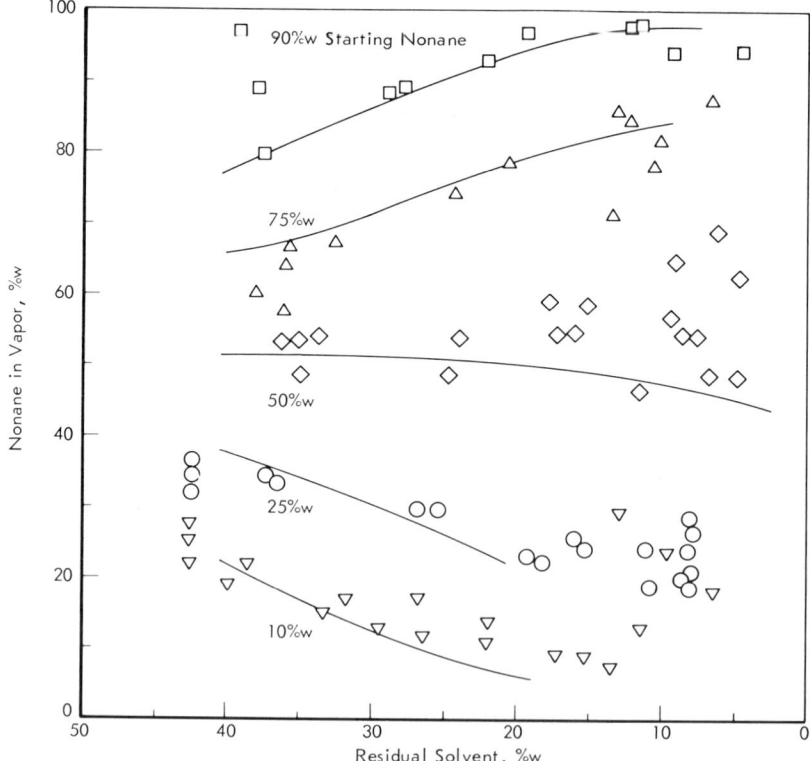

Figure 15. Vaporization of nonane–BuOH blend from Beckosol 34 resin

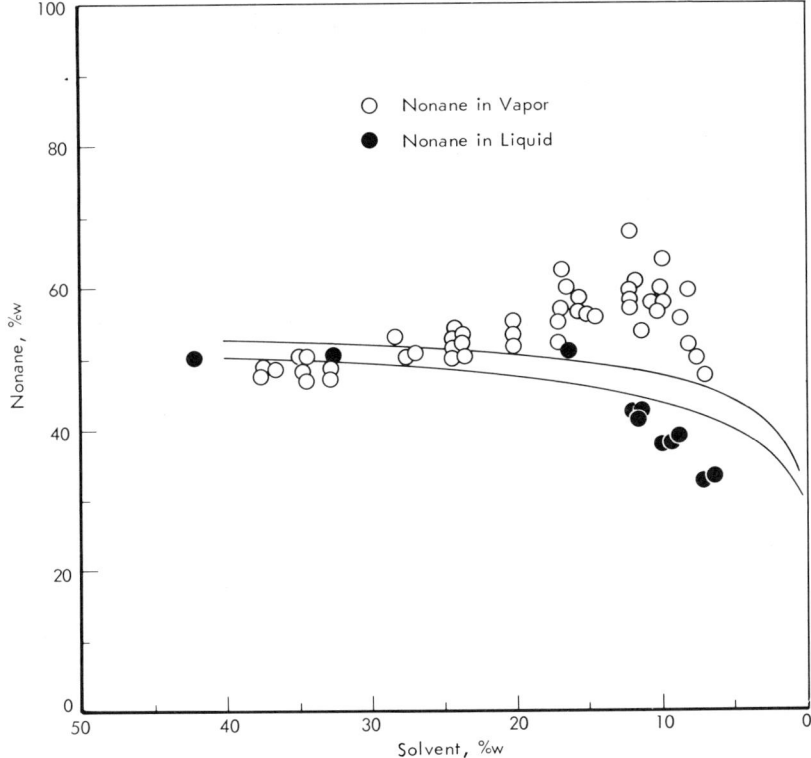

Figure 16. Vaporization of nonane–BuOH blend from Beckosol 34 resin

Notation

P_i	Partial pressure component i
y_i	Vapor mole fraction
x_i	Liquid mole fraction
γ_i	Activity coefficient
a_i	Activity
v_{FHi}	Number of size or FH groups in molecule i
R_i	Ratio size molecule i to average
v_{ki}	Number of k groups in molecule i
Γ_k	Activity coefficient of k group in solution
Γ_k^i	Activity coefficient of k group in pure i
X_k	Group fraction of k groups
a_{lk}, a_{kl}	Group pair interaction parameters for groups l and k
Σ_l	Sum over all groups
Σ_i	Sum over all molecules
wt %	Weight fraction
χ	Flory interaction parameter

Literature Cited

1. Hansen, C. M., *Ind. Eng. Chem. Prod. Res. Develop.* (1970) **9**, 282.
2. Larson, E. C., Sletmoe, G. M., *Gravure*, August–September (1964).
3. Sletmoe, G. M., *J. Paint Technol.* (1966) **38**, 642.
4. Sletmoe, G. M., *J. Paint Technol.* (1970) **42**, 246.
5. Derr, E. L., Deal, C. H., *Proc. Intern. Symp. Distill., Brighton, Engl.* (1969) No. **32**, 3:40.
6. Deal, C. H., Derr, E. L., *Ind. Eng. Chem.* (1968) **60**, 28.
7. Nelson, R. C., Hemwall, R. W., Edwards, G. D., *J. Paint Technol.* (1970) **42**, 636.
8. Nelson, R. C., Figurelli, V. F., Walsham, J. C., Edward, G. D., *J. Paint Technol.* (1970) **42**, 644.
9. Walsham, J. G., Edwards, G. D., *J. Paint Technol.* (1971) **43**, 64.
10. Heil, J. F., Prausnitz, J. M., *AIChE J.* (1966) **12**, 678.
11. Pierotti, G. J., Deal, C. H., Derr, E. L., *Ind. Eng. Chem.* (1959) **51**, 95.
12. Hillyer, M. J., Leonard, W. J., Advan. Chem. Ser. (1973) **120**, 31.

RECEIVED March 14, 1972.

Calculation of Concentrated Polymer Solution Viscosities: A New Approach

M. J. HILLYER[a] and W. J. LEONARD, Jr.[b]

Shell Development Co., P. O. Box 24225, Oakland, Calif. 94623

> *A model for calculating viscosities of concentrated polymer solutions has been formulated and used successfully to predict viscosities of alkyd resin solutions in both pure aromatic solvents and in mixtures of hydrocarbons and oxygenated materials. It was also found to describe viscosity trends in polystyrene–diethylbenzene solutions accurately. The formulation explicitly accounts for the observation that concentrated solution viscosities increase markedly with decreasing compatibility between resin and diluent. The proposal of an empirical relationship which interprets the viscosity enhancement in poorer solvents in terms of increased chain-chain interactions is of interest. The model contains three constants which are fixed for a particular resin and are independent of diluent type. These are the Mark-Houwink constant, the parameter in the Martin viscosity equation, and the constant relating the postulated clustering to the solution thermodynamics of a particular solution.*

The viscosities of concentrated polymer solutions exhibit a solvent dependency which differs markedly from that observed for dilute solutions. Measurements on resins in a variety of solvents at resin concentrations above 30 wt % show that solution viscosities increase with decreasing solvent power. This trend cannot be explained by solvent perturbation effects on chain dimensions since this leads to the prediction that, *ceteris paribus*, viscosities in good solvents are higher than in poor solvents.

---

[a] Present address: Shell Development Co., P. O. Box 481, Houston, Texas 77001.
[b] Present address: Dynapol, 1454 Page Mill Road, Palo Alto, Calif. 94304.

Doolittle (1) has examined the mechanism of solvent action in concentrated resin solutions and has suggested a mechanistic picture which qualitatively accounts for the observed solvent effects. In this model two equilibria are considered; aggregation-disaggregation and solvation-desolvation. The resin molecules are viewed as being in a state of continuous flux between solvation and aggregate formation. Clearly, poor solvents promote aggregation, and good solvents favor solvation. Since the aggregates formed have a larger hydrodynamic volume than the individual chains, the poor solvents which favor their formation will give solutions with higher viscosities.

Various theories have been proposed which take into account the effects of aggregation on the viscosities of polymer solutions (2). These theories generally require a detailed knowledge of the mechanism and energetics of association. Further, they are considered applicable only to moderately concentrated solutions.

The objective of the present work is to formulate a model capable of quantitatively accounting for the solvent effects on viscosity. Our approach is empirical. It involves calculating the effective molecular weight of the resin as a function of concentration and the thermodynamic properties of the solution. The latter are estimated using the analytical solution of groups (ASOG) method of Deal and Derr (3). The model is formulated on the premise that the hydrodynamically effective molecular weight obtained in poor solvents can be much larger than in good solvents because of the increased intermolecular segmental interactions of the resin. It follows that the larger hydrodynamic units enhance the solution viscosities.

The model has been tested for its ability to predict Newtonian solution viscosities at polymer concentrations in the range of 30–60 wt % in a number of pure solvents and solvent blends. The calculated results generally agree with the experimental data.

The Model

An expression relating the effective hydrodynamic molecular weight M_{eff} of the resin to its true molecular weight, M, must account for two facts: 1, the probability of a chain segment interacting with another segment of a different chain is proportional to the resin concentration in solution, and 2, the probability of segment-segment interaction is greater when the mixing free energy is less favorable (*i.e.*, less negative). An expression which incorporates both of these observations is

$$M_{eff} = M \exp\left(\phi_1^{-\xi}\gamma_1 - 1\right) \qquad (1)$$

where ϕ_1 is the volume fraction of solvent, ξ is a constant and γ_1 is the volume activity coefficient of the solvent. The latter is defined as

$$\gamma_1 = a_1/\phi_1$$

with a_1 being the solvent activity of the mixture. The parameter ξ is specified to be independent of the solvent. Its value should be in the range $\xi \geq 0$.

Equation 1 was formulated after examining a large number of resin viscosity and solvent activity data. The data were available for a number of resin systems and covered a broad range of solvent type. Various empirical expressions were screened for their ability to match trends in viscosity with solvent activity coefficients. Equation 1 consistently gave effective molecular weights which, at a given concentration, corresponded well with the observed shifts in viscosity. This equation is used in conjunction with an appropriate viscosity equation to extend its use to a range of concentration.

Equation 1 is used to calculate the effective intrinsic viscosity $[\eta]_c$ of the resin using the Mark-Houwink equation

$$[\eta]_c = KM^\alpha_{eff} \text{ where } \alpha = 1/2 \tag{2}$$

As used here $[\eta]_c$ denotes the increment in relative viscosity which results when a resin molecule is added to the solution at concentration c. Consequently, its value for a given resin can strongly depend on the concentration, and on the nature of the diluent. At infinite dilution Equation 1 gives $M_{eff} = M$, and the intrinsic viscosity has its conventional meaning, i.e., the limiting value of $(\eta_s - \eta_0)/\eta_0 c$ at infinite dilution. Here η_s is the viscosity of a solution of concentration c (grams/dl) and η_0 is the solvent viscosity.

Theory (4) shows that the constant K in the Mark-Houwink equation should be independent of the molecular weight and solvent. However, the value $1/2$ for the molecular weight exponent α in Equation 2 is strictly applicable to systems in which the solvent is a θ-solvent. A larger value for α is generally observed for solvents which perturb the chain configuration. These effects can be significant for high molecular weight polymers but appear to be of decreasing importance as the molecular weight decreases. Recent studies show that many polymers with molecular weights below about 10^4 have α values equal to $1/2$ (5). Since the resins of interest here have molecular weights well below this limit, an exponent of $1/2$ seems applicable.

The solution viscosity is calculated using the Martin equation

$$\eta_s = \eta_0[1 + [\eta]_c c \exp(k[\eta]_c c)] \tag{3}$$

In this equation, k is a constant which usually depends very little on the solvent type. In its present application all solvent effects are postulated to be manifested in the value of $[\eta]_c$ via Equation 1. Agreement between theory and experiment, demonstrated in the next section, shows that this assumption is valid for the systems studied.

Results and Discussion

Concentrated Resin Solutions. The model has been tested on three short oil-alkyd resins in a number of solvent systems. The resins studied are Allied Chemical Co.'s resin No. 2918, and Reichhold Chemical Co.'s Beckosol 3 and Beckosol 7. Chemical and physical properties of these resins which are used in the viscosity calculations are given in Table I. The chemical group compositions are given in Table II as moles of each group per 100 grams of resin. The group counts as listed are required input data for the ASOG activity calculations (3).

Table I. Properties of Short Oil Alkyd Resins

Property	*Allied 2918*	*Beckosol 3*	*Beckosol 7*
Acid number	7–11	26–36	6–14
Phthalic anhydride, %	48	44	42
Oil, %	33	33	41
Oil type	Coconut	Soya	Soya
Resin	None	Present	None
Phenolic resin	None	None	None
Specific gravity	1.19	1.18	1.16
Molecular Weight [a]	2370	1550	2120

[a] Ebullioscopic, in benzene.

Table II. Group Compositions of Short Oil Alkyd Resins (Moles per 100 Grams)

Group	*Allied 2918*	*Beckosol 3*	*Beckosol 7*
—C(=O)—	.7986	.7448	.7400
—O—	.7986	.7448	.7400
—OH	.0336	.0308	.0294
(phenyl ring)	.3456	.3168	.3024
—HC=CH—	.2244	.2312	.2788
—CH$_2$—	2.2506	2.2288	2.5052

The ASOG method generates molecular or volume activity coefficients for each component in a mixture from a matrix of known pair interactions between each type of chemical group. Using the group counts in Table II and similar counts on the diluent, all permutations for interaction are calculated and properly weighted to give the solution activity coefficients.

Table III. Viscosities of Short Oil Alkyd Resins in Aromatic Solvents[a]

Solvent	Concentration, Wt. Fraction Resin	Viscosity, cSt		
		Allied 2918	Beckosol 3	Beckosol 7
Xylene	0.20	4.73	3.32	3.32
	0.30	21.3	13.5	15.0
	0.40	160.	107.	76.3
	0.50	866.	975.	717.
	0.60	5870.	10500.	5390.
Cumene	0.20	5.85	4.09	3.8
	0.30	35.0	18.2	14.4
	0.40	304.	191.	127.
	0.50	1983.	2860.	1680.
	0.60	15300.	34600.	13000.
n-Butylbenzene	0.20	8.80	5.48	5.24
	0.30	59.3	29.1	21.1
	0.40	751.	363.	184.
	0.50	6470.	8180.	2900.
tert-Butylbenzene	0.20	10.7	6.16	5.46
	0.30	77.6	32.3	24.3
	0.40	870.	414.	247.
	0.50	7750.	8560.	3830.

[a] Data made available by E. C. Larson, Shell Development Co., Oakland, Calif.

The viscosity model was first tested for its ability to describe the solution viscosities of the three resins over a broad concentration range in a number of pure solvents. The viscosity data used are given in Table III. The ASOG solvent activities for each binary mixture are given as a function of concentration in Figures 1, 2, and 3. The activity curves for n-butylbenzene and tert-butylbenzene for each resin are identical since the ASOG method does not differentiate between molecules having the same group compositions. In all cases the curves predict a phase separation at solvent volume fractions roughly in the range 0.2–0.6 and indicated by $a_1 > 1.0$.

The phase separations predicted by the ASOG calculations appear to be inconsistent with the experimental observations on these binarys. The solutions have the appearance of being homogeneous. On the other hand, molecular aggregation without precipitation is anticipated by the model presented here and strongly suggested by the observations of

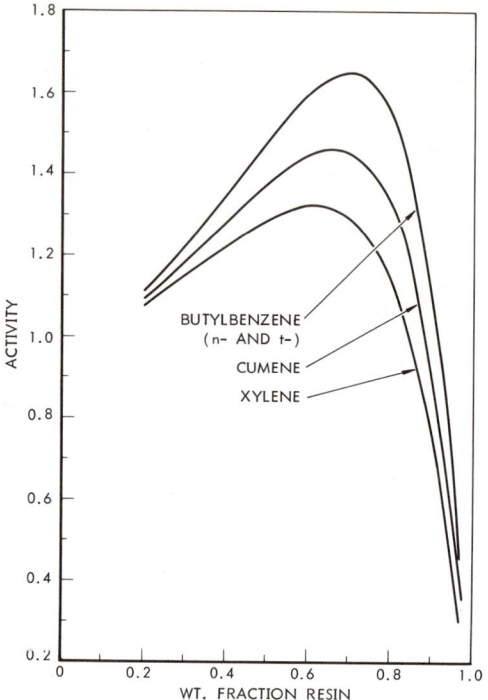

Figure 1. Solvent activity by ASOG for Allied 2918 resin

others (1, 6). The complex compositions of the resins, as shown in Table I, may account for stabilization of the solution. In particular, the oil component may act as a suspending agent and, as such, have a stabilizing influence on the resin aggregates or clusters.

If this interpretation of the thermodynamics is correct, the solvent activities should be invariant at all concentrations within the critical region. The exact location of this region is given by zero equivalence of the first and second derivatives of the chemical potential ($\log a_1$) with respect to concentration. We have chosen to define this region, arbitrarily, as the region where solvent activity exceeds unity since the tedious calculation necessary to establish it exactly seemed unwarranted by the approximate nature of the model. The maximum resin concentration at which the activity initially exceeds unity is defined as the "critical concentration." For the three systems under study, these concentrations are listed in Table IV. It follows that γ_1 in Equation 1 is given by $1/\phi_{crit}$, where ϕ_{crit} is the critical solvent volume fraction.

The model was then fitted to the data in Table III, using a nonlinear estimation technique attributable to Meyer (7). The calculated model parameters for the three resins are given in Table V. The degree

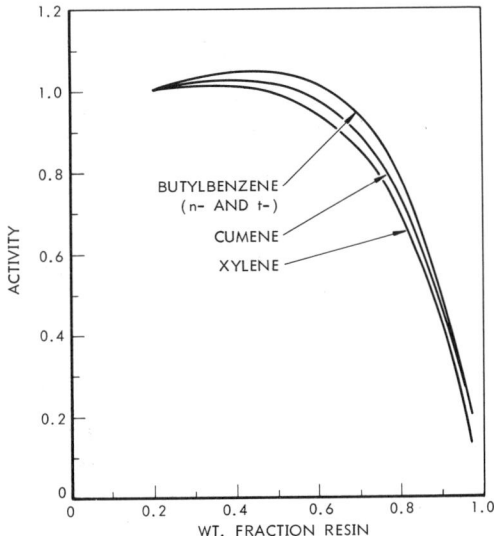

Figure 2. Solvent activity by ASOG for Beckosol 3 resin

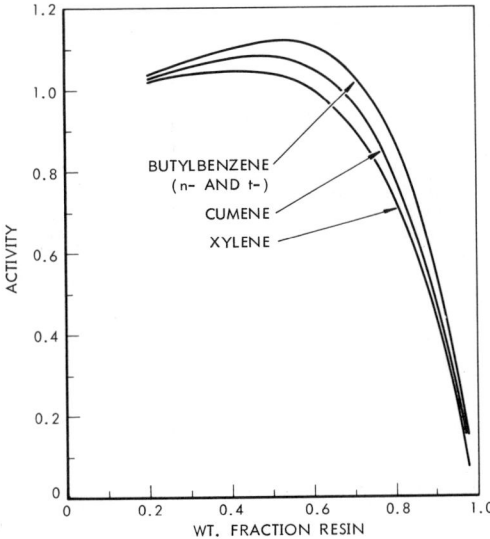

Figure 3. Solvent activity by ASOG for Beckosol 7 resin

Table IV. Critical Solvent Concentrations for Alkyd Resins

Resin	Solvent Volume Fraction		
	Xylene	Cumene	Butylbenzene
Allied 2918	0.197	0.150	0.114
Beckosol 3	0.608	0.523	0.447
Beckosol 7	0.467	0.394	0.324

Table V. Model Parameters for Short Oil Alkyd Resins

Resin	Parameter		
	K	ξ	k
Allied 2918	1.093×10^{-3}	0.1038	1.882
Beckosol 3	5.447×10^{-4}	0.3954	5.209
Beckosol 7	4.132×10^{-4}	0.2589	5.844

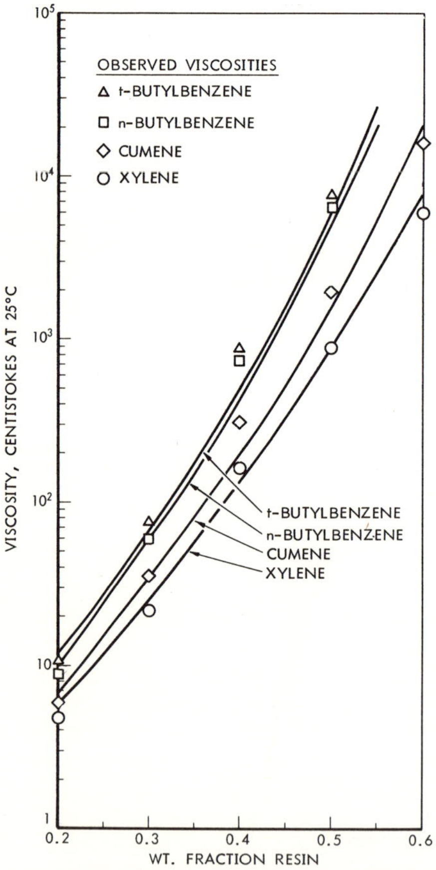

Figure 4. Viscosities of Allied 2918 resin solutions. Smooth curves represent calculated viscosities.

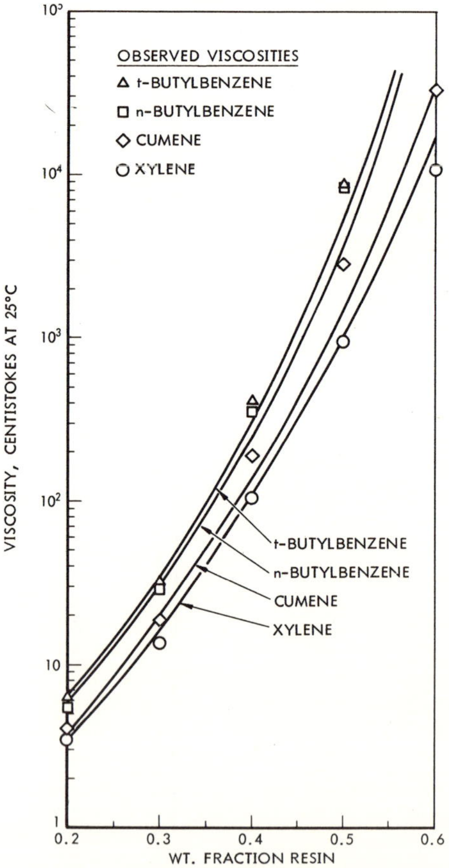

Figure 5. Viscosities of Beckosol 3 resin solutions. Smooth curves represent calculated viscosities.

of fit of the model to the data may be judged from Figures 4, 5, and 6, where the solution viscosities are plotted against concentration. The calculated curves agree with the experiment to a remarkable degree, considering that the data cover a range of viscosity of four orders of magnitude. The values of the parameter K are of the same order of magnitude as found in other works. For example, Flory (4) reports values of K in the range of 10^{-3} for a number of typical polymer systems. The Martin constants k are somewhat larger than expected. A theoretical analysis indicates these constants should range approximately between 0.3 and 2.0 (2). The discrepancy may either be caused by differences in the types of systems considered, or it may be an artifact of our model.

Having obtained the resin parameters it is possible to calculate the apparent degree of association or aggregation, M_{eff}/M, using Equation 1. The results for the Allied 2918 resin in the solvents used in the above

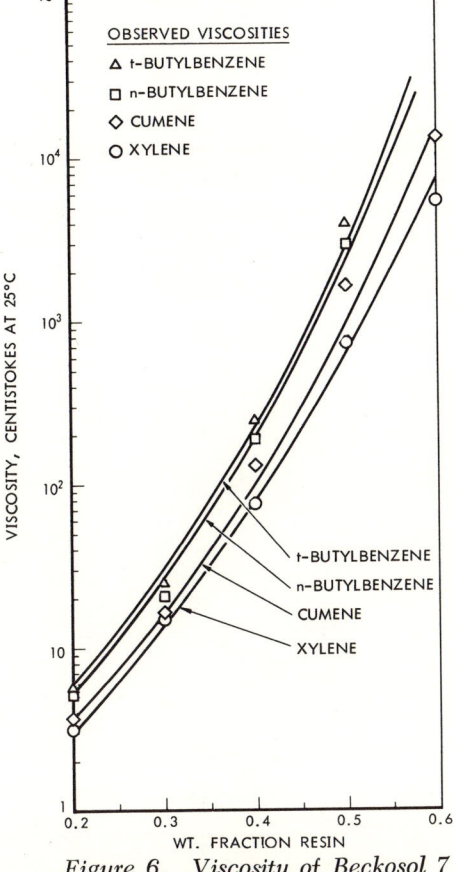

Figure 6. Viscosity of Beckosol 7 resin solutions. Smooth curves represent calculated viscosities.

viscosity studies are shown in Figure 7 as a function of resin volume fraction $\phi_2 = 1 - \phi_1$. The curves illustrate the large effects expected from relatively small differences in solvent type, especially at high resin concentrations. The values of M_{eff}/M, as calculated, are subject to considerable uncertainty. Small differences in the viscosity data and calculated activity coefficients will change the levels of apparent association, as will changes in the model, *per se*. On the other hand, the calculations do support the concept of aggregation in resin solutions.

Large differences in solvent power effect large changes in solution viscosity. Data presented by Reynolds (8) show that Beckosol 7 viscosities increase sharply on going from a solvent with a solubility parameter matching the resin (9.7) to one a few units lower. Similar behavior is predicted by the present model, as shown in Figure 8. The solubility parameters are related to the activity coefficients by (9)

$$\gamma = \exp\left[\frac{V_1}{RT}(\delta_1 - \delta_2)^2\right] \quad (4)$$

where V_1 is the molar volume of the solvent, δ_1 and δ_2 are the solubility parameters of the solvent and resin respectively, and RT has its usual meaning. For the calculations presented, the ratio V_1/RT was assigned the value 1/6.

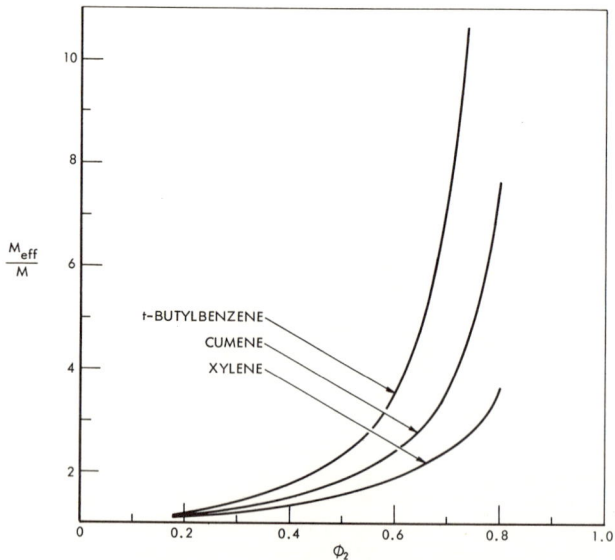

Figure 7. Calculated aggregation of Allied 2918 resin

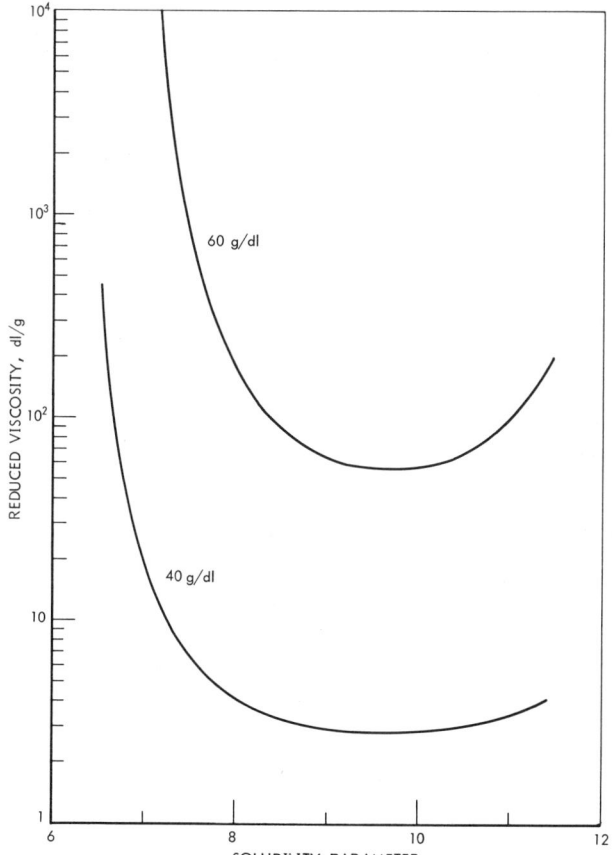

Figure 8. Reduced viscosities of Beckosol 7 as predicted using model coefficients and a range of solubility parameters

With the resin parameters evaluated, the practical utility of the model was tested as follows. Data on mixed solvents containing both hydrocarbons and oxygenated materials were used to assess the ability of the model to describe the viscosity of systems different from those which determined the parameters. The compositions of the solvents are given in Table VI. For the ASOG calculations, the group compositions of the mineral spirits were calculated from the chemical composition data of these solvents. The solution viscosity was calculated using the model as specified in Equations 1–3. However, the solvent blend is treated as a single liquid with an average activity coefficient given by

$$\gamma_1 = \exp\left(\sum_i \phi_i' \ln \gamma_i\right) \qquad (5)$$

Table VI. Composition of

Blend Component	1	2	3	4	5	6	7	8	9	10	11	12
VM&P naphtha #1	50							50	50			
VM&P naphtha #2		50	50	50	50	50	50			65	60	50
Mineral spirits (High boiling, Narrow range)												
Mineral spirits (Low boiling, Narrow range)												
Mineral spirits (Low boiling, Isoparaffinic)												
Mineral spirits (High boiling, Isoparaffinic)												
n-Butyl acetate		20	20					20	20			29
n-Butyl alcohol		19		39				19	39	22	27	
Oxitol	19		19			19						
Oxitol acetate												
Butyl oxitol												
Dioxitol												
Cyclohexanone												10
2-Nitropropane			39									
sec-Butyl alcohol					39							
Methyl isobutyl ketone							5					
Toluene										13	13	
Ethylbenzene	11	11	11	11	11	11	6	11	11			11
Mesityl oxide												
Cyclohexane												

where γ_i and ϕ_i' are the activity coefficient and volume fraction of each component in the solvent blend. Similarly, the individual solvent viscosities $\eta_{0,i}$ are averaged by

$$\eta_0 = \exp \left(\sum_i \phi_i' \ln \eta_{0,i}\right) \quad (6)$$

The calculated mixed solvent viscosities, together with the measured viscosities, are given in Table VII. The viscosities are given in terms of the Gardner-Holdt scale. Figure 9 shows the same data as Table VII. The agreement between theory and experiment is remarkably good especially in view of the polar nature of some of the solvents.

Concentrated Polystyrene Solutions. We also applied our model to the data of Bueche (*10*) in the polystyrene–diethylbenzene system. These

Mixed Solvents (Volume %)

13	14	15	16	17	18	19	20	21	22	23	24	25	26	27	28
50	50	50	60	50	50	50									
							20	23	40	24	24	17	20	24	40
							20	17		16	16	23	20	16	
							24	36		36	36	28	28	28	
									30						30
20			34	39	20										
					19										
	39	9							8	16		12	24		
						28	16	20	8		24	12			6
															24
				11											
11	11	11	11	11		8	8			8	8	8	8	8	
19	39		31	5				10							

data cover polymer concentrations up to 75 grams/dl for polymers having molecular weights of 5×10^4 to above 10^6. The parameters for Equations 1–3 were established on the basis of the viscosity data for the polymer of molecular weight 5.4×10^5. Values of γ_1 for the system were estimated from the Flory-Huggins theory by the equation

$$\gamma_1 = \exp(\phi_2 - \chi\phi^2_2) \qquad (7)$$

The interaction parameter, χ, was assigned the value 0.434 (*11*). An excellent fit was obtained when the model was applied to the data, which covered a concentration range of 20–75 grams/dl. These results are shown in Figure 10. The fit may be judged from the average deviation of the predicted viscosities from the observed viscosities, which was 20%

Table VII. Observed and Calculated Viscosities for Alkyd Resins in Mixed Solvents

Resin	Solvent Blend	Wt. Fraction Resin	Observed Viscosity	Calculated Viscosity
Allied 2918	1	0.6	Z1–Z2	Z3–Z4
	2	0.6	Z2	Z3+
	3	0.6	Z1–Z2	Z3–Z4
	4	0.6	X	Z3–Z4
	5	0.5	W–X	V
	6	0.5	V–W	V–W
	7	0.6	Z2–Z3	Z3–Z4
	8	0.6	M–N	Z3
	9	0.6	Z3	Z4–Z5
	10	0.4	L	E
	10	0.6	Z5	Z3+
	11	0.4	I–J	E–F
	12	0.6	Z3–Z4	Z2+
	13	0.6	Z2–Z3	Z1–Z2
	14	0.6	Z1–Z2	Z1–Z2
	15	0.6	Z6	Z3–Z4
	16	0.6	Z5	Z2
	17	0.6	Z4–Z5	Z1–Z2
	18	0.6	Z4–Z5	Z1–Z2
	19	0.5	U	U+
	19	0.6	Z2+	Z3+
Beckosol 7	20	0.5	W	W
	21	0.5	Z1–Z2	U–V
	22	0.5	Y	X+
	23	0.5	Z3	U+
	24	0.5	Z4	U
	25	0.5	W–X	V–W
	26	0.5	X	V+
	27	0.5	Y	U–V
	28	0.5	X	Z1

for the eight results spanning a viscosity range of seven orders of magnitude. The model parameters were evaluated as $\xi = 0.166$, $K = 2.5 \times 10^{-3}$, $\alpha = 0.5$, and $k = 0.077$.

The molecular weight dependence of viscosity can be calculated by combining equations 1–3 and taking the partial derivative

$$\frac{\partial \ln \eta}{\partial \ln M} = \alpha \left[\frac{\eta - \eta_0}{\eta} \right] [1 + kcKM^{\alpha}_{eff}] \qquad (8)$$

where the symbols have their previous meanings. At high polymer concentrations $\eta \gg \eta_0$. Thus,

$$\frac{\partial \ln \eta}{\partial \ln M} \simeq \alpha [1 + kcKM^{\alpha}_{eff}]. \qquad (9)$$

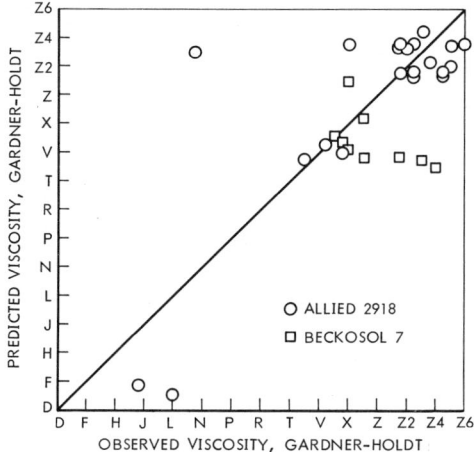

Figure 9. Observed and predicted Gardner–Holdt viscosities in mixed solvents

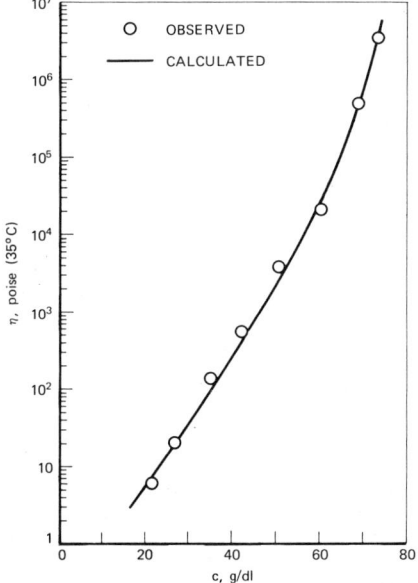

Figure 10. Viscosity of polystyrene–diethylbenzene solutions

The derivative ($\partial \ln \eta / \partial \ln M$) is identified with the exponent in the empirical correlation $\eta \simeq M^\beta$ which is frequently applied to concentrated polymer solutions (*12*).

Using the parameters obtained above on the polymer with molecular weight 5.4×10^5, values of β were then calculated for a range of molecular weights. The value of c was set at 41 grams/dl, which is the concentration at which Bueche (*10*) reported viscosity data for many polymers with different molecular weights. The values calculated for β increase monotonically from about 1.5 at molecular weights between 10^4 and 10^5, to about 4.0 for polymers with molecular weights around 10^6. Analysis of Bueche's reported experimental data shows excellent agreement with these calculated values at both ends of this molecular weight range. The deviation in the center of the range is about 20%.

Conclusions

A new approach to calculating viscosities of concentrated polymer solutions has been presented. It consists of the derivation of a semi-empirical computational model containing three parameters characteristic of a particular polymer. Once these parameters have been established, the viscosity of any solution of the polymeric material in a solvent or solvent blend may be calculated. The method should be of particular interest to the coatings industry, where they often require a screening estimate of the potential viscosity-reducing power of a new solvent blend.

A notable feature of the model is the postulation of an expression relating aggregation of polymer chains to concentration and the solvent power of the diluent. The excellent agreement between prediction and observation for solutions of short oil alkyd resins is taken as evidence for the importance of aggregation effects on viscosity in these systems. In addition, the accurate description of the molecular weight dependence of the viscosities of concentrated polystyrene solutions is an indication of the general applicability of the model.

Literature Cited

1. Doolittle, A. K., "The Technology of Solvents and Plasticizers," Chapter 14, John Wiley and Sons, New York, 1954.
2. Frisch, H. L., Simha, R., in "Rheology," Vol. I, F. R. Eirich, Ed., p. 525, Academic, New York, 1956.
3. Deal, C. H., Derr, E. L., *Proc. Intern. Symp. Distill., Brighton, Engl.* (1969) No. 32, 3:40.
4. Flory, P. J., "Principles of Polymer Chemistry," Cornell University Press, Ithaca, 1953.
5. Bianchi, U., Peterlin, A., *J. Polym. Sci., Pt. A-2* (1968) **6**, 1759.
6. Dintenfass, L., *J. Oil Colour Chem. Ass.* (1957) **40**, 761.
7. Meyer, R. R., in "Non-Linear Programming," J. B. Rosen, O. L. Mangasarian, K. Ritter, Eds., p. 465, Academic, New York, 1970.
8. Reynolds, W. W., "The Physical Chemistry of Petroleum Solvents," Reinhold, New York, 1963.

9. Hildebrand, J. H., Scott, R. L., "Regular Solutions," Prentice-Hall, Englewood Cliffs, N. J., 1962.
10. Bueche, F., *J. Appl. Phys.* (1953) **24**, 423.
11. Boyer, R. F., Simha, R., in "Styrene, Its Polymers, Copolymers and Derivatives," R. H. Boundy, R. F. Boyer, Eds., p. 314, Reinhold, New York, 1952.
12. Fox, T. G., Gratch, S., Loshaek, S., in "Rheology," F. R. Eirich, Ed., Vol. I, p. 431, Academic, New York, 1956.

RECEIVED March 14, 1972.

4

Solvent Selection by Computer

CHARLES M. HANSEN

PPG Industries, Inc., 151 Colfax St., Springdale, Pa. 15144

A linear programming technique is described which selects mixed solvents based on specifications of the δ_D, δ_P, and δ_H solubility parameters, evaporation rates, and other significant parameters of a solvent blend. Suggestions are made for setting the various restrictions and for setting procedures of data processing. For simpler cases of solvent selection, recourse to a computer is not necessary. Use of a solvent improvement cost factor, δ_a/cost, then leads to optimum formulations since one can determine which solvents increase solubility at least cost. δ_a is given by $\sqrt{\delta_P^2 + \delta_H^2}$.

The present discussion extends the method of linear programming reported by Nelson *et al.* (1) as an aid in solvent selection. This extension uses the most accurate system available to predict the type of solubility, the δ_D, δ_P, and δ_H solubility parameters (2, 3, 4, 5, 6, 7) and contains refinements incorporated over the years this procedure has been in use. This approach to solubility prediction has its origin in Hildebrand's work (8, 9) as extended by Blanks and Prausnitz (10). Other systems for predicting solubility (11, 12, 13, 14, 15, 16, 17) may be useful for similar purposes, but their accuracy is not as great (7). Situations such as reformulation of solvents to meet air pollution codes are particularly well suited for this procedure as well as simple cost reduction.

The solubility parameter system used here assumes that the energy of evaporation—*i.e.*, the total cohesive energy which holds a liquid together, E—can be divided into contributions from three forces: (1) dispersion (London) forces, E_D, (2) permanent dipole–permanent dipole forces, E_P, and (3) hydrogen-bonding forces, E_H. Thus,

$$E = E_D + E_P + E_H$$

Dividing this equation by the molar volume of a solvent, V, gives:

$$\frac{E}{V} = \frac{E_D}{V} + \frac{E_P}{V} + \frac{E_H}{V}$$

or

$$\delta^2 = \delta^2_D + \delta^2_P + \delta^2_H$$

where $\delta = (E/V)^{1/2} = $ (cohesive energy density)$^{1/2}$
This is the usual equation for the solubility parameter (8).

$$\delta_D = (E_D/V)^{1/2}$$

is the dispersion (London) component of the solubility parameter.

$$\delta_P = (E_P/V)^{1/2}$$

is the polar component of the solubility parameter.

$$\delta_H = (E_H/V)^{1/2}$$

is the hydrogen bonding component of the solubility parameter.

Permanent dipole–induced dipole forces are generally small compared with the other forces with all errors being largely collected in the hydrogen bonding component. The units for all these parameters are (cal/cc)$^{1/2}$.

Solubility data can be plotted in a three-dimensional system using the δ_D, δ_P, and δ_H parameters as coordinates. The solubility regions for various polymers will be spherical if the δ_D axis is expanded with a unit distance equal to twice that of the unit distances on the other axes. This effect is attributed to the directional nature of the molecular forces involved in δ_P and δ_H vs. the omnidirectional nature of the atomic dispersion forces. Description of solubility in terms of a geometrically simple model has many advantages, particularly in computer processing.

Linear Programming

Linear programming requires linear equations to calculate average properties of a solvent blend from the properties of the individual solvents and their respective (volume) fractional amounts. Thus each average property of the blend, P_N, is calculated from the properties of the individual pure components, P_{N_i}, by an equation of the form:

$$\overline{P}_N = \sum_{i=1}^{j} v_i P_{N_i} \tag{1}$$

Here v_i is the volume fraction of the "i"th component of the blend, and the summation is over all of the j solvents present. The various $\overline{P}_N$ may be evaporation rate, solubility parameters, content of aromatic compounds, or whatever the formulator desires to control in his formulation. The results will be as accurate as the ability of the linear equations to represent the true average properties.

The linear programming technique locates that combination of solvents which provides the specified properties at minimum cost. One may require that a given $\overline{P}_N$ be equal to, greater than, or less than a given value with one equation. More flexibility is possible when a range on the $\overline{P}_N$ is allowed by combined use of the greater-than and less-than equations.

Since there are many different linear program packages and since a typical example of computer input to such a package already exists in the literature (1), this discussion is concerned with the choice of restrictions and relates practical experience gained from using the method described. This program led to a tabulation of solvent improvement costs, and reference to this tabulation can occasionally lead to solution of a problem without recourse to a computer.

General Procedure

The solvent selection procedure described here has been designed for use by plant laboratories on a time sharing terminal. To eliminate the need for excessive schooling, the computer input has been set up as a form of conversation with the computer. The operator responds to questions by specifying the solvents to be considered, which $\overline{P}_N$ he wishes to specify, and the limits within which the final $\overline{P}_N$ are to be restricted. This input is treated within the program to set up an appropriate file for input of information to the actual linear program. Solvent property data are stored in a permanent file which is serviced by a maintenance program for updating costs or for adding new solvents. This file is also accessed by other programs dealing with solvents.

The results of the linear program optimization are placed in a file which serves as input to a program for computing the average properties of the resulting optimized blend. This program serves several functions. It conveys the properties and cost of the final answer in tabulated form and also prints a δ_P vs. δ_H solubility parameter plot of the individual solvents, the blend, and any polymers which may be of interest. It is often used to define the properties of a non-Rule 66 blend so that this can be matched with a Rule 66 solvent blend. (Rule 66-type legislation, which originated in Los Angeles County to restrict allowable solvent compositions in coatings, has been picked up in other areas with local variations in limits.) It can also be used to define the averaged properties of any fixed solvent in the total formulation (that is, solvents which appear in the resin additions). The fixed solvent data can be temporarily entered into the solvent file and treated as a single solvent which must be present at a given concentration. This procedure speeds up the computing process.

A typical reformulation example would include the following steps:

(1) Determination of the total solvent and fixed solvent in the system.

(2) Computer evaluation of the $\overline{P_N}$ for the total solvent and the fixed solvent.

(3) Linear program determination of a solvent to yield the $\overline{P_N}$ desired.

(4) Computer evaluation of the properties of the suggested blend.

(5) Practical evaluation of the suggested blend recognizing that slight modifications are often desirable to perfect the system properties.

Solvency Criteria

The $\delta_D - \delta_P - \delta_H$ system is exceptionally accurate for the solubility predictions desired in this procedure (2, 3, 4, 5, 6, 7). The solubility regions for the polymers can be expressed as spheres (3, 4). Where two or more polymers are involved, the over-lapping region of solubility guides the formulation. A problem arises, however, in trying to describe a sphere with linear equations. A satisfactory compromise for defining the region of solubility is to use a cube having the same center as the sphere and with sides equal in length to 0.8 the diameter of the sphere.

The computer requests the input of the center and radius of the sphere and sets the upper and lower bounds on each of the three solubility parameters. This means six equations are required to specify the region of solubility.

Where a solvent is to be matched, it is not necessary to know the coordinates of the region of solubility for the polymer. One can usually surmise that the optimum solution will have a maximum of (relatively inexpensive) hydrocarbon, but simply requiring that the matching solvent have all three solubility parameters greater than the standard will not yield an optimum solution, if indeed, it will yield a feasible result at all. Restriction of the δ_P and δ_H parameters to be greater than the standard is usually necessary, but the δ_D parameter may be greater than or less than the standard in most cases. This is particularly true where aromatic hydrocarbons are to be replaced since these systems have generally had more solvency than is needed. Since many polymers have similar regions of solubility, one method of fulfilling these recommendations is to use a fictitious region of solubility with a radius of four units and a center placed such that the location of the solvent to be matched is at the lower allowable limits of the δ_P and δ_H parameters. The δ_D coordinate of the sphere is the same as that of the solvent to be matched. Should an improvement in a solvent blend be desired, one can simply enter a smaller radius.

A problem can arise if a hydrocarbon non-solvent having an evaporation rate lower than that of the average solvent blend is included in the formulation. To eliminate this possibility of eventual precipitation of the polymer, one should not consider these solvents as potential ingredients in the formula or else specify a given amount of still less volatile, good solvent.

Evaporation Rate

The properties of the optimized solvent blend generally meet the restrictions of minimum solvency and either a lower or upper limit on average evaporation rate. The use of two equations to place upper and lower bounds on the evaporation rate is recommended. If the evaporation rates are too restrictive, the computer tends to select solvents which are more expensive and of better quality. It is possible to set such stringent or unreasonable restrictions that no solution is acceptable to the computer. It is instructive at times to eliminate the evaporation rate restrictions altogether; evaporation rate can be controlled by the solvents the computer is allowed to consider. Our solvent selection programs have operated satisfactorily with the relative evaporation rate system based on *n*-butyl acetate as 100 although this is an arbitrary choice which has not been evaluated by comparison.

Other Restrictions

A necessary restriction in the linear programming input is the requirement that the volume fractions of the final blend components add up to 1.0. Other helpful restrictions are that minimum amounts of given solvents be present in the formulation, or conversely, that upper limits be set on the concentrations of given solvents. A Rule 66-type formulation could be handled in this manner, but a more satisfactory manner is to use three equations placing maximum restrictions on the three classes as specified by the Rule. For example, Class I solvents must be less than 0.05 volume fraction, Class II solvents must be less than 0.08 volume fraction, and the sum of the three classes must be less than 0.20 volume fraction. As Nelson (*1*) discussed, the viscosity may be controlled although we have not found this to be a significant factor, and we often omit the specification. In this case the logarithms of the viscosities of the individual solvents are additive by the volume fractions rather than the viscosities themselves.

Another factor which may become increasingly important is the flash point. Unfortunately there is no simple relation to estimate flash points of mixtures, so this limits the accuracy of flash paint restrictions in the linear programming technique. We have included an option to maintain the flash point above a given value. Since the additivity is based on simple volume fraction additivity of the flash points, this option can only be considered a very rough guide.

Discussion

The usefulness of the linear programming technique lies in its ability to consider simultaneously many factors and arrive at a result to satisfy the needs of the given situation. If the end result is consistently regulated by a single factor, there is no need for the program. Linear programming technique is very useful in Rule 66-type formulations where, for example, a number of hydrocarbon solvents may be possible in a formula and each one has its respective solvent restrictions. This also points out the need to know the exact levels of Rule 66 classes in the solvents as supplied to the final user. Only with this information can an optimum formulation be found, since the maximum levels of aromatic solvents allowed are usually found in Rule 66-type solvents.

The procedure used in the linear program itself can be used to establish a table of modified solvent costs allowing optimization without resorting to the computer. A table of δ_A/cost for non-hydrocarbon solvents, where δ_A is defined as:

$$\delta_A = \sqrt{\delta^2_P + \delta^2_H} = \sqrt{\delta^2 - \delta^2_D}$$

shows which solvents increase the solvency of a hydrocarbon-rich system at the least cost per unit of solubility parameter (δ_A) improvement. The δ_P vs. δ_H plot is suitable for most solvent selection problems (6). The δ_D parameter is relatively constant and need not be considered specifically in many practical situations. If the combined solvency and cost data are arranged in order from most-likely to least-likely candidates, one can start at the top of the list and proceed downward until a suitable solvent is found that meets the approximate requirements of evaporation rate, odor, or whatever restrictions may be necessary. Lower molecular weight solvents stand at the top of the list as can be seen in Tables I, II, and III which contain the cost factor for various solvents.

One word of caution is needed here: δ_A cannot be used indiscriminately for every polymer. It is well known that ketones are excellent solvents for epoxy resins while higher δ_A solvents like alcohols are not.

Table I. δ_A/Cost for High Evaporation Rate Solvents

Solvent	δ_A $(Cal/cm^3)^{1/2}$	Bulk Cost[a] $/Gal	$\delta_A/Cost$
Methanol	12.4	0.12	103.3
2-Propanol	8.56	0.47	18.2
Ethanol	10.45	0.565	18.5
Acetone	6.13	0.37	15.57
Methyl ethyl ketone	5.06	0.68	7.44
Ethyl acetate	5.19	0.89	5.83
Isopropyl acetate	4.57	0.84	5.44

[a] Data from *Oil, Paint and Drug Reporter* (Nov. 20, 1972).

Table II. δ_A/Cost for Medium Evaporation Rate Solvents

Solvent	δ_A $(Cal/cm^3)^{1/2}$	Bulk Cost[a] $/Gal	$\delta_A/Cost$
2-Butanol	8.30	0.53	15.66
Butanol	8.20	0.74	11.1
2-Nitropropane	6.23	1.32	4.72
Isobutyl acetate	4.57	0.87	5.3
Methyl isobutyl ketone	4.15	0.90	4.6
n-Butyl acetate	3.58	0.98	3.65

[a] Data from *Oil, Paint, and Drug Reporter* (Nov. 20, 1972).

Table III. δ_A/Cost for Low Evaporation Rate Solvents

Solvent[b]	δ_A $(Cal/cm^3)^{1/2}$	Bulk Cost[a] $/Gal	$\delta_A/Cost$
EGMEE	8.55	1.34	6.4
Diacetone alcohol	6.60	1.10	6.0
EGMBE	6.67	1.56	4.28
EGMEEA	5.63	1.36	3.89
Isophorone	5.37	1.38	3.89
Cyclohexanone	4.79	1.41	3.40

[a] Data from *Oil, Paint, and Drug Reporter* (Nov. 20, 1972).
[b] EGMEE = ethylene glycol monoethyl ether.
EGMBE = ethylene glycol monobutyl ether.
EGMEEA = ethylene glycol monoethyl ether acetate.

The same is true of poly(vinyl chloride). In a similar manner high δ_A solvents are poor for non-polar polymers such as butyl rubber. The δ_A/cost factor is most useful for replacing aromatic solvents, for example, by quickly pointing to the use of isobutyl alcohol or n-butyl alcohol in conjunction with an aliphatic hydrocarbon to replace as much xylene as may be necessary to meet air pollution regulations. Computer processing will lead to the same suggestion. In any event the recommended blend requires laboratory evaluation prior to use commercially whether it is derived from a computer recommendation or from a δ_P vs. δ_H plot using the δ_A/cost parameter to suggest which solvents should be evaluated.

Conclusion

Experience has shown that the procedures outlined above are effective. They can be modified, and the number of restrictive equations can be changed as desired. The linear programming technique requires a computer for full benefit, and the time sharing system is preferable because of the direct contact between formulator and computer. Without a computer, use of a δ_P vs. δ_H plot together with the modified cost described above and a table of solvent properties should enable a reasonably rapid and economically satisfactory result.

Acknowledgment

The author thanks Wesley French in developing the solvent selection procedure described above. Thanks are also due to PPG Industries, Inc. for permission to publish this work.

Literature Cited

1. Nelson, R. C., Figurelli, V. F., Walsham, J. G., Edwards, G. D., *J. Paint Technol.* (1970) **42**, 644.
2. Hansen, C. M., *J. Paint Technol.* (1967) **39**, 104.
3. Hansen, C. M., *J. Paint Technol.* (1967) **39**, 505.
4. Hansen, C. M., Skaarup, K., *J. Paint Technol.* (1967) **39**, 511.
5. Hansen, C. M., *Ind. Eng. Chem. Res. Dev.* (1969) **8**, 2.
6. Hansen, C. M., Beerbower, A., "Encyclopedia of Chemical Technology Supplement Volume," Chapter on "Solubility Parameters," p. 889, 1971.
7. Beerbower, A., Dickey, J. R., *Amer. Soc. Lubrication Engrs. Trans.* (1969) **12**, 1.
8. Hildebrand, J., Scott, R., "Solubility of Non-Electrolytes," 3rd ed., Reinhold, New York, 1949.
9. Hildebrand, J., Scott, R., "Regular Solutions," Prentice-Hall, Englewood Cliffs, N. J., 1962.
10. Blanks, R. F., Prausnitz, J. M., *Ind. Eng. Chem. Fundam.* (1964) **3**, 1.
11. Burrell, H., *Off. Dig. Fed. Soc. Paint Technol.* (1955) **27**, 726.
12. Burrell, H., *J. Paint Technol.* (1968) **40**, 197.
13. Lieberman, E. P., *Off. Dig. Fed. Soc. Paint Technol.* (1962) **34**, 30.
14. Dyck, M., Hoyer, P., *Farbe Lack* (1964) **70**, 522.
15. Crowley, J. D., Teague, G. S., Lowe, J. W., *J. Paint Technol.* (1966) **38**, 296.
16. Crowley, J. D., Teague, G. S., Lowe, J. W., *J. Paint Technol.* (1967) **39**, 19.
17. Gardon, J. L., *J. Paint Technol.* (1966) **38**, 43.

RECEIVED March 14, 1972.

5

Prediction of Flash Points for Solvent Mixtures

JOHN G. WALSHAM

Shell Development Co., P. O. Box 24225, Oakland, Calif. 94623

> *Flash points of mixtures of oxygenated and hydrocarbon solvents cannot be predicted simply. A computer based method is proposed which exhibits satisfactory prediction of such Tag Open Cup flash points. Individual solvent flash point indexes are defined as an inverse function of the component's heat of combustion and vapor pressure at its flash point. Mixture flash points are then computed by trial and error as the temperature at which the sum of weighted component indexes equals 1.0. Solution non-idealities are accounted for by component activity coefficients calculated by a multicomponent extension of the Van Laar equations. Flash points predicted by the proposed method are compared with experimental data for 60 solvent mixtures. Confidence limits of 95% for differences between experimental and predicted flash points are +8.0–+3.0°F.*

A recently developed computer-based technique allows time spent in laboratory screening of solvent formulations for particular applications to be greatly reduced (1). In this method, blend performance properties are calculated by simple mixing rules from the properties of the component solvents, and a blend composition is chosen which will match a required set of properties while minimizing cost. Solvent blends formulated in this way have an excellent chance of success in subsequent laboratory tests.

Flash point is an important solvent blend performance property. Mixture flash points cannot be predicted using linear mixing rules, and to date no literature describing a satisfactory method of predicting flash

Present address: 1801 Jefferson St., San Francisco, Calif. 94123.

points for mixtures of oxygenated and hydrocarbon solvents has been found. Such a method would be helpful in developing solvent blends to meet particular performance, labeling, and safety requirements. This paper describes an approach to calculating solvent blend flash points which further extends the capability of computer screening of solvent blend formulations.

The flash point of a flammable liquid is an experimentally determined ignition temperature and may vary considerably depending on the type of apparatus used in its determination. Several ASTM-approved methods are available for determining the flash points of hydrocarbon fractions and of pure organic liquids (2).

The flash point most often quoted for oxygenated solvents is the Tagliabue (Tag) Open Cup or TOC flash point while for hydrocarbons the Tag Closed Cup or TCC flash point is more commonly used. When hydrocarbon and oxygenated solvents are mixed together, the TOC flash point is generally the one chosen to characterize the blend. The main concern here will be that of predicting TOC flash points for solvent blends from the flash points of the individual components.

The flash point of a multicomponent system cannot be determined by summing a simple fraction of the flash points of the individual components. If, however, the component flash points were defined by some temperature dependent property which might then be calculated for a multicomponent system, the flash point of that system might be predicted.

An early approach to predicting TCC flash points for hydrocarbon mixtures used this concept by estimating flash point as the temperature at which mixture vapor pressure reached about 10 mm Hg (3). Other workers (4) have found both the product of vapor pressure at the flash point and molecular weight to be constant for a number of hydrocarbon fractions. The TCC flash point for hydrocarbon blends was predicted as the temperature at which

$$\sum_{i=1}^{N} x_i P_i = 15.2 \tag{1}$$

where x_i = mole fraction of component i in the blend, P_i = vapor pressure of component i, mm Hg, and N = number of components in the blend.

A further approach to estimating flash points of mixed hydrocarbons is to sum volume fraction weighted flashing indexes which are calculated directly from the flash points of the individual components (5). Blend flash points estimated by the latter two methods generally agreed with experimental values within the limits of reproducibility of experimental data.

For hydrocarbon fractions there are established correlations between TCC flash point and boiling range (6, 7). In the present work a review of TOC flash point data for oxygenated solvents shows a relatively good correlation (standard deviation of 5%) between log (component vapor pressure at 25°C) and flash point. Correlation between flash point and boiling point proved to be less satisfactory.

Vapor pressure at the flash point and the product of molecular weight and vapor pressure at the flash point were computed for 40 oxygenated solvents. Vapor pressure at the flash point varied between 7 and 58 mm Hg while the product of vapor pressure and molecular weight varied between 800 and 4200. Thus, some of the above methods of predicting flash points of mixed hydrocarbons appear unlikely to be effective for systems containing oxygenated solvents with their attendant non-ideal behavior.

Prediction of TOC Flash Points for Mixed Solvents

If, for an individual solvent, a flashing index is defined as

$$I_i = 1/T_i$$

where T_i is a function of some temperature-dependent property of an individual solvent, T_i takes its value at the flash point of that component, then the flash point of an N-component mixture can be calculated as the temperature at which

$$\sum_{i=1}^{N} I_i x_i T_i = 1.0 \qquad (2)$$

where x_i is the mole fraction of component i in the mixture.

If values of T_i can be predicted for each component at various temperatures, a trial-and-error procedure might be used to converge on the mixture flash point.

As a first approach to utilizing this flashing index calculation technique, component flashing indexes were defined as the reciprocal of the component vapor pressure at the flash point. Mixture flash points were then computed by trial-and-error as the temperature at which

$$\sum_{i=1}^{N} I_i x_i P_i \gamma_i = 1.0 \qquad (3)$$

where P_i is the vapor pressure of component i at the trial temperature and γ_i equals activity coefficient of component i in the mixture (*see* Appendix 1).

A computer-based file of solvent property data was used in conjunction with this procedure. Thus, component vapor pressures at various temperatures were generated by interpolation from file-stored data on component boiling points and vapor pressures at 25°C.

Mixture flash point predictions using this definition of flashing index were not substantially more accurate than the commonly used approximation of assuming a mixture flash point to be equal to that of the lowest flashing component. This simple definition of flashing index assumes that the flash point of a flammable solvent is determined only by its vapor pressure. It is evident, however, when pure component flash point data are reviewed, that within a group of similar compounds, vapor pressure at the flash point decreases with increasing component molecular weight. This trend might be expected since flash point is a temperature at which ignition can be sustained in the vapor phase and probably is therefore related to molar heat of combustion. In any homologous series, heat of combustion is almost a linear function of molecular weight. When a liquid mixture evaporates, the vapor becomes more concentrated in the lighter components. However, the mean molecular weight of the vapor, and hence its mean heat of combustion, are higher than that of the individual light components present, so the observed flash point might be expected to fall below the determined flash point using Equation 2.

Regression analysis of data on 25 common solvents (both hydrocarbon and oxygenated) yielded the relationship:

$$\text{Heat of Combustion} \, \alpha \, (\text{Molecular Weight})^{1.25} \qquad (4)$$

and flashing index was redefined as:

$$I_i = \frac{1}{PF_i M_i^{1.25}} \qquad (5)$$

where PF_i is the vapor pressure of component i at its flash point, and M_i is the molecular weight of component i. For a mixture of N components, the flash point is predicted as the temperature at which:

$$\sum_{i=1}^{N} I_i x_i P_i \gamma_i M_v^{1.25} = 1.0 \qquad (6)$$

where M_v = calculated mean vapor molecular weight at the test temperature.

Inclusion of the mean vapor molecular weight in Equation 6 recognizes that this represents the environment in which the flash will occur. The mean molecular weight of the vapor at the trial temperature is calculated by the computed component vapor pressures and activity coefficients.

When non-combustible components are present in a mixture, their presence in the vapor space does not contribute to the support of combustion at the flash point. In such cases, M_v is calculated as the mean molecular weight of only the flammable components present in the vapor phase. This approach of course ignores the likely effects of different non-combustible components on mixture flash points but nevertheless provides an effective model.

The first column of Table I lists TOC flash points for 58 solvent blends; these data were retrieved partly from open literature (8) and partly from earlier laboratory reports circulated within Shell Chemical Co. TOC flash points for 38 of these blends were redetermined, and these data are listed in the second column of Table I. For each of the total of 60 solvent blends referred to in Table I, the flash point of the lowest flashing component in the blend and the blend flash point computed using the sum of flashing indexes method of Equations 5 and 6 are also listed. The flash point of the lowest flashing component in a solvent blend is often used in practice as an estimate of the blend flash point. The composition of the 60 blends is given in Table II along with the TOC flash points of the constituent solvents. These single solvent flash points were the ones used in the flashing index calculation procedure.

Comparison of Calculated and Experimental TOC Flash Points

Flash points calculated by the proposed sum-of-indexes method are compared in Table III with experimentally determined flash points. To place this comparison in context, Table III also includes several other comparisons between sets of data. When earlier experimentally determined data were compared with flash points redetermined for this study, the differences were almost evenly distributed around the zero point. The median absolute difference between the two sets of data was 5°F while the largest difference between flash points determined for a single blend was 27°F. Some skew is apparent in the differences between either set of experimental data and the calculated flash points. The median absolute differences between experimental and calculated data were either 7° or 8°F depending upon which set of experimental data was used in the comparison. When the flash point of the lowest flashing component was taken as an indication of blend flash point, the median differences from the two sets of experimental data were 13° and 18°F, respectively.

For the two sets of experimental data, the 95% confidence limits for differences between flash point determinations were $+3.2°$ to $-3.2°F$.

Table I. Solvent Blend TOC Flash Points: Comparison Between Published Flash Points, Redetermined Flash Points, and Flash Points Calculated by the Sum of Indexes Method[a]

Blend No.	Published Data TOC Flash Point, °F[b]	Redetermined TOC Flash Points, °F	Lowest Flashing Component TOC Flash Point, °F	Calculated TOC Flash Point, °F
1	42	—	48	49
2	35	—	34	33
3	32	—	34	31
4	33	—	20	25
5	40	—	48	41
6	38	28	41	46
7	40	—	48	48
8	80	—	89	83
9	77	102	88	91
10	54	—	49	49
11	36	—	34	41
12	31	46	34	47
13	27	—	34	40
14	65	—	60	65
15	83	84	60	77
16	67	62	20	49
17	14	12	19	29
18	93	100	80	87
19	76	64	48	53
20	71	64	60	62
21	70	66	20	61
22	74	74	49	63
23	85	92	67	81
24	89	82	67	76
25	67	70	49	63
26	72	72	49	59
27	69	70	49	54
28	68	74	49	71
29	84	76	67	75
30	81	—	88	89
31	99	92	88	88
32	90	—	88	88
33	98	—	100	101
34	88	—	80	85
35	113	134	116	121
36	94	96	67	93
37	68	—	55	58
38	67	—	55	58
39	68	—	55	59
40	67	—	55	59
41	69	—	55	60
42	60	54	20	41
43	71	62	20	59

Table I. Continued

Blend No.	Published Data TOC Flash Point, °F [b]	Redetermined TOC Flash Points, °F	Lowest Flashing Component TOC Flash Point, °F	Calculated TOC Flash Point, °F
44	72	74	60	67
45	61	56	48	48
46	66	58	48	51
47	41	36	20	33
48	44	42	20	38
49	46	50	55	57
50	59	32	−20	14
51	56	32	−20	12
52	76	58	20	52
53	67.5	66	60	68
54	68	70	60	73
55	70.5	68	60	75
56	73	74	60	75
57	74.5	80	60	74
58	81.5	82	60	73
59	—	82	67	79
60	—	66	48	57

[a] Composition of the solvent blends is given in Table III.
[b] For data on blends 1–9 and 12–13 *see* Probst (8).

For the calculated data the corresponding limits were +8.0° to 1.8°F (earlier experimental *vs.* calculated flash points) and +8.0° to +3.0°F (redetermined experimental *vs.* calculated flash points). These limits further indicate the skew in the calculated data and show that a conservative, or low, flash point prediction is more likely than a high prediction.

Discussion

The accuracy of predicting mixture flash points depends strongly on the validity of the pure component flash point data which are used in the calculation procedure. There is, for example, little published data for TOC flash points of hydrocarbons. Some of the accepted TOC flash point data for oxygenated solvents were unreliable, *e.g.*, the TOC flash point acetone is often quoted as 15°F; this was redetermined in the present work at −20°F. Component flash points shown in Table III were generally taken from published flash point data and were not checked experimentally.

The proposed calculation procedure is effective in handling blends of a wide range of industrial solvents, but it has proved to be less successful in predicting flash points for systems with extremely volatile components. It is difficult to obtain consistent TOC data for these systems because of the rapid vaporization of the low boiling materials. Additionally, if the volatile components are flammable, there are few data on their pure component flash points while the effect of non-flammable volatile components in excluding oxygen from the liquid surface is not accounted for in the current procedure.

The U. S. Department of Transportation has announced (9) that future labeling regulations for flammable liquids may be based on their TCC flash points. If this change is made, the empirical calculation technique proposed here is expected to be equally applicable to predicting TCC data if the closed cup flash points of the individual solvents are substituted in basic calculation data. No attempts have yet been made to predict closed cup data, but one might expect that the vapor space in the closed cup would be more easily simulated than that in the open cup tester.

Table II. Composition of Solvent Blends Referred To in Table I

Blend No.	Components	Composition, % v	Component Flash Points TOC, (°F^a)
1	Toluene	73	48
	n-Butyl alcohol	27	106
2	Ethyl acetate	77	34
	Isopropyl alcohol	23	60
3	Ethyl acetate	57	34
	Methyl alcohol	49	55
4	Isopropyl alcohol	30	60
	Methyl ethyl ketone	70	20
5	Isopropyl alcohol	58	60
	Toluene	42	48
6	Methyl alcohol	72	55
	Toluene	28	48
7	Toluene	70	48
	Methyl isobutyl carbinol	30	—
8	Xylene	70	89
	Methyl isobutyl carbinol	30	106
9	n-Butyl acetate	50	88
	n-Butyl alcohol	50	114
10	Toluene	66	49
	n-Butyl alcohol	34	106
11	Methyl alcohol	82	55
	Ethyl acetate	18	34
12	Ethyl acetate	33	34
	Isopropyl alcohol	33	60
	Cyclohexanone	34	129
13	Toluene	48	48
	Isopropyl alcohol	33	60

Table II. Continued

Blend No.	Components	Composition, % v	Component Flash Points TOC, (°F a)
	Ethyl acetate	20	34
	n-Butyl acetate	9	88
14	Isopropyl alcohol	75.5	60
	n-Butyl alcohol	24.5	106
15	Isopropyl alcohol	34.6	60
	n-Propyl alcohol	33.9	90
	n-Butyl alcohol	16.9	106
	Oxitol glycol ether	14.6	130
16	Methyl ethyl ketone	13	20
	sec-Butyl alcohol	32	80
	Diacetone alcohol	22.1	136
	Isopropyl alcohol	32.9	60
17	Super VM&P naphtha	23	24
	Isopropyl alcohol	13	60
	Acetone	11	−20
	Methyl isobutyl ketone	12	67
	Toluene	31	48
	Isobutyl alcohol	8.5	100
	n-Butyl alcohol	1.5	106
18	Diacetone alcohol	50	136
	sec-Butyl alcohol	50	80
19	Toluene	72.7	48
	Pentoxone solvent	27.3	141
20	Methyl isobutyl ketone	69.8	67
	Isopropyl alcohol	12.2	60
	Isobutyl alcohol	18.0	100
21	Pentoxone solvent	5	141
	Methyl isobutyl ketone	90	67
	Methyl ethyl ketone	5	20
22	Methyl isobutyl ketone	41	67
	Toluene	26	48
	Ethyl amyl ketone	14	120
	Ethylene glycol monoethyl ether acetate	19	150
23	Methyl isobutyl ketone	48.5	67
	Ethyl amyl ketone	24	120
	Ethylene glycol monoethyl ether acetate	10	150
	Xylene	17.5	89
24	Pentoxone solvent	30	141
	Methyl isobutyl ketone	70	67
25	Methyl isobutyl ketone	75	67
	Toluene	10	48
	Xylene	15	89
26	Methyl isobutyl ketone	62	67
	Toluene	21.5	48
	Ethyl amyl ketone	13.5	120
27	Methyl isobutyl ketone	47	67

Table II. Continued

Blend No.	Components	Composition, % v	Component Flash Points TOC, (°F[a])
	Toluene	43.5	49
	Ethylene glycol monoethyl ether acetate	4.9	150
	Butyl Oxitol glycol ether	4.6	165
28	Methyl isobutyl ketone	54	67
	Toluene	10	48
	Xylene	4	89
	Neosol solvent	4	58
	Methyl amyl acetate	21	100
	Methyl isobutyl carbinol	7	114
29	Methyl isobutyl ketone	46.7	67
	Butyl Oxitol glycol ether	1.1	165
	n-Butyl alcohol	5.7	106
	n-Butyl acetate	46.5	88
30	n-Butyl acetate	40.2	88
	Isobutyl acetate	59.8	90
31	n-Butyl acetate	85	88
	Isobutyl acetate	15	90
32	n-Butyl acetate	75	88
	Isobutyl acetate	25	90
33	n-Butyl alcohol	20	106
	Isobutyl alcohol	80	100
34	Methyl isobutyl carbinol	31	114
	sec-Butyl alcohol	69	80
35	Ethyl amyl ketone	75	120
	Cyclo Sol 53	25	124
36	Methyl isobutyl ketone	5	67
	Ethyl amyl ketone	52	120
	Cyclo Sol 53	33	124
	sec-Butyl alcohol	10	80
37	Methyl alcohol	75	55
	Isopropyl alcohol	22.5	60
	Water	2.5	—
38	Methyl alcohol	70	55
	Isopropyl alcohol	27	60
	Water	3	—
39	Methyl alcohol	65	55
	Isopropyl alcohol	31.5	60
	Water	3.5	—
40	Methyl alcohol	60	55
	Isopropyl alcohol	36	60
	Water	4	—
41	Methyl alcohol	55	55
	Isopropyl alcohol	40.5	60
	Water	4.5	—
42	Isopropyl alcohol	9.8	60
	Toluene	51	48

Table II. Continued

Blend No.	Components	Composition, % v	Component Flash Points TOC, (°F a)
	Methyl ethyl ketone	9.8	20
	Methyl isobutyl ketone	25.6	67
	Diisobutyl ketone	3.8	120
43	Isopropyl alcohol	17	60
	Xylene	48.4	89
	Methyl ethyl ketone	6.7	20
	Methyl isobutyl carbinol	16.5	114
	Diacetone alcohol	11.4	136
44	Isopropyl alcohol	25	60
	Xylene	25	89
	Diacetone alcohol	50	136
45	Methyl isobutyl ketone	35	67
	Toluene	50	48
	Isopropyl alcohol	6	60
	Methyl isobutyl carbinol	9	114
46	Methyl isobutyl ketone	43	67
	Toluene	57	48
47	Toluene	41.4	48
	Isopropyl alcohol	15.3	60
	Methyl isobutyl ketone	15.4	67
	Methyl amyl acetate	1.5	100
	Tolu Sol 5	15	20
	Xylene	8.4	84
	Methyl ethyl ketone	3	20
48	Isopropyl alcohol	20	60
	Methyl isobutyl ketone	25	67
	Toluene	42	48
	Tolu Sol 5	8	20
	Oxitol glycol ether	5	130
49	Isopropyl alcohol	50	60
	Methyl alcohol	50	55
50	1,1,1-Trichloroethane	84.4	—
	Acetone	15.6	−20
51	1,1,1-Trichloroethane	83.3	—
	Acetone	16.7	−20
52	Pentoxone solvent	72.7	141
	Methyl ethyl ketone	27.3	20
53	Isopropyl alcohol	72.7	60
	Water	27.3	—
54	Isopropyl alcohol	91	60
	Water	9	—
55	Isopropyl alcohol	80	60
	Water	20	—
56	Isopropyl alcohol	70	60
	Water	30	—
57	Isopropyl alcohol	60	60
	Water	40	—

Table II. Continued

Blend No.	Components	Composition, % v	Component Flash Points TOC, (°F^a)
58	Isopropyl alcohol	50	60
	Water	50	—
59	Methyl isobutyl ketone	65	67
	Methlene chloride	35	—
60	Toluene	75	48
	Methylene chloride	25	—

^a Component flash points shown here are those used in calculating blend points appearing in Table I.

Table III. Comparisons Between Sets of TOC Flash Point Data From Different Sources

Data Being Compared with Earlier Experimentally Determined Flash Points	No. of Data Points	Differences between Data, °F		
		Mean	Median Absolute	95% Confidence Limits
1. New experimentally determined flash points	38	1.76	5	+3.2 to −3.2
2. Flash points calculated by sum of indexes method	58	4.88	8	+8.0 to +1.8
3. Flash points of lowest flashing component in the solvent blend	58	14.7	13	+19.8 to +9.7
Data Being Compared with New Experimentally Determined Flash Points				
1. Flash points calculated by sum of indexes method	40	5.25	7	+8.0 to +3.0
2. Flash points of lowest flashing component in the solvent blend	40	19.2	18	+25.5 to +13.0

Flash points of systems containing dissolved solids have not been considered in the present treatment. A more universal model would be necessary to quantify solvent–solute interactions and to predict flash point modification by solutes. Such a model would be valuable in screening total solvent–solute formulations.

Appendix I. Calculation of Component Activity Coefficients

Component activity coefficients are calculated by a multicomponent extension of the Van Laar equations. Development of this method has been described elsewhere (*10*), and only a summary is given here.

In binary mixtures the Van Laar equations allow the component activity coefficient predictions at any composition by interpolation from the activity coefficient values at infinite dilution γ_{12}^∞, γ_{21}^∞ of components 1 in 2 and 2 in 1, respectively. In the multicomponent extension, component activity coefficients at any mixture composition are computed by interpolation of binary system infinite dilution activity coefficients from a matrix.

The general relationship for the activity coefficient of component i in a system of n components is:

$$\frac{\log \gamma_i}{S_i} = \sum_{j=1}^{n} \phi_j B_{ij} - \sum_{j=1}^{n} \sum_{k>j}^{n} \phi_j \phi_k B_{kj} \qquad (7)$$

Here the weighted fraction of any component i, φ_i, is given by

$$\phi_i = \frac{x_i S_i}{\sum_{j=1}^{n} x_j S_j} \qquad (8)$$

The weighting factors S_i are derived from the binary limiting activity coefficients by representing these in the form:

$$\log \gamma_{ij}^\alpha = S_i B_{ij} \qquad \log \gamma_{ji}^\alpha = S_j B_{ij} \qquad (9)$$

S_i and S_j are properties of the individual components i and j, B_{ij} is an interaction parameter characteristic of the binary system ij, and B_{ii} is equal to zero. For any multicomponent system, allowing the weighting factor S_i of one component to be unity enables calculation of weighting factors and interaction parameters for all components and binary combinations in the mixture.

The general equation for component activity coefficients is derived from summing the binary interactions throughout a multicomponent system. For example, the activity coefficient for component three of a ternary mixture is given by

$$\frac{\log \gamma_3}{S_3} = \phi_1 B_{13} + \phi_2 B_{23} - \phi_1 \phi_2 B_{12} - \phi_1 \phi_3 B_{13} - \phi_2 \phi_3 B_{23} \qquad (10)$$

The present application requires one to calculate activity coefficients in blends containing components drawn from a list of more than 100 solvents. Compiling the individual limiting activity coefficients for all binary systems in this list would be impractical. The approach has been to divide the solvents into groups according to their structural characteristics and to use a matrix showing only interactions between the solvent groups. The relevant limiting activity coefficients between solvent groups are shown in Table IV.

Table IV. Matrix of Limiting Binary Activity Coefficients between Component Types[a,b]

Type	1	2	3	4	5	6	7	8	9
1	1.0	1.0	12.7	33.9	2.9	2.7	3.7	2.0	48.0
2	1.0	1.0	15.0	57.0	2.8	2.2	3.7	2.0	28.0
3	5.0	5.2	1.0	1.3	1.0	1.2	2.0	1.3	83000.0
4	6.8	25.0	1.4	1.0	2.6	2.8	7.4	4.0	300000.0
5	3.0	2.1	1.2	2.8	1.0	1.0	1.6	1.1	1200.0
6	2.5	2.0	1.5	4.3	1.0	1.0	0.7	1.3	180.0
7	13.0	4.4	1.0	3.6	1.0	1.0	1.0	1.5	519.0
8	3.1	1.8	1.3	8.8	1.1	0.9	0.7	1.0	350.0
9	3.8	2.0	575.0	1000.0	8.3	12.0	16.0	2.4	1.0

[a] Components are divided into the following nine classes. The single component taken as representative of its class in deriving the above data is shown in parentheses.
    Type 1. Alcohols (n-Butyl alcohol)
           2. Ether alcohols (Ethylene glycol monoethyl ether)
           3. Aromatic hydrocarbons (Cumene)
           4. Aliphatic hydrocarbons (Nonane)
           5. Esters (n-Butyl acetate)
           6. Ketones (Methyl isobutyl ketone)
           7. Chlorinated aliphatic hydrocarbons (Dichloroethane)
           8. Glycol ether esters (Ethylene glycol monomethyl ether acetate)
           9. Water

[b] The following illustrations show the order of indices to be used in reading activity coefficients from the table:

$$\gamma^\alpha_{1,4} = 33.9; \quad \gamma^\alpha_{4,1} = 6.8$$

Acknowledgment

The author thanks T. E. Mounsey for determining the new flash point data reported in this paper.

Literature Cited

1. Nelson, R. C., et al., *J. Paint Technol.* (1970) **42**, 644.
2. 1967 Book of ASTM Standards, **Part 20**, ASTM Designations No. 93, 56, and 1310.
3. Thiele, E. W., *Ind. Eng. Chem.* (1927) **19**, 259.
4. Butler, R. M., et al., *Ind. Eng. Chem.* (1956) **48**, 808.
5. Wickey, R. O., Chittenden, D. H., *Hydrocarbon Process. Petrol. Refiner* (1963) **42**, 157.
6. Nelson, W. L., *Oil and Gas J.*, 63 (Aug. 8, 1966).
7. Burger, L. L., *U.S. At. Energy Comm.* Publication **4W-35579** (1955).
8. Probst, K. G., et al., *J. Paint Technol.* (1969) **41**, 670.
9. *Fed. Regist.* (1970) **35**, (236) 18534.
10. Black, C., *Ind. Eng. Chem.* (1958) **50**, 391, 403.

RECEIVED March 14, 1972.

6

The Photochemical Smog Reactivity of Organic Solvents

ARTHUR LEVY

Battelle, Columbus Laboratories, Columbus, Ohio 43201

> *The chemical solvent industry is greatly affected by current air-pollution regulations on local, state, and national levels. Current control laws are based in large part on Los Angeles' Rule 66 which ranks specific solvents and classes of solvents according to their potential smog-forming capability. A number of studies have been conducted over the past several years that are aimed at defining the relationship between organic solvents and photochemical-smog production. This paper reviews and analyzes some of the more recent of these. Solvents are ranked according to various photochemical smog parameters, taking into account many of the conditions (differences in chambers, concentration, light intensity, etc.) under which the particular studies were conducted.*

When one discusses organic solvents and air pollution, there is an almost Pavlovian response to Los Angeles' Rule 66. In some respects it is unfortunate that this response has developed as rapidly as it has and to the extent that it has across the nation. When the rule was promulgated, Lewis J. Fuller who was the air pollution control officer of Los Angeles County, cautioned those who attempted to apply this rule to the rest of the country. He said that other communities should first learn the character and extent of their local problems and then try to resolve these problems, possibly taking advantage of some aspects of Rule 66. He also pointed out that the transfer of Rule 66 to other areas might stretch the utility of the rule too far.

In the mid-1960's when Rule 66 was being promulgated, the problem of photochemical smog was technologically still in its infancy. We are now much more aware of the complexities of the atmospheric chemistry that are associated with the reactions of hydrocarbons and nitrogen oxides in the presence of sunlight. As a result, although an extensive amount of work was conducted before the rule was promulgated, it has become increasingly evident that the definition of reactivity of species in the atmosphere is an extremely complex problem—one that is associated with so many parameters that it is difficult to extrapolate basic laboratory-type reactivity experiments.

Experimental reactivity is a problem unto itself, being a function of chamber design, chamber operation, and parameters such as hydrocarbon–nitric oxide concentration, humidity, background gas impurity levels, and light intensity. For a detailed discussion of this area of smog reactivity and for more general reviews of photochemical smog reactivity, the reader is referred to Refs. *1, 2,* and *3.* In this paper some of the studies which followed the Los Angeles Air Pollution Control District (*4*) studies are reviewed, principally to show some of the complexities that exist and some results that may modify and help other agencies to apply the reactivities of organic solvents to atmospheric conditions.

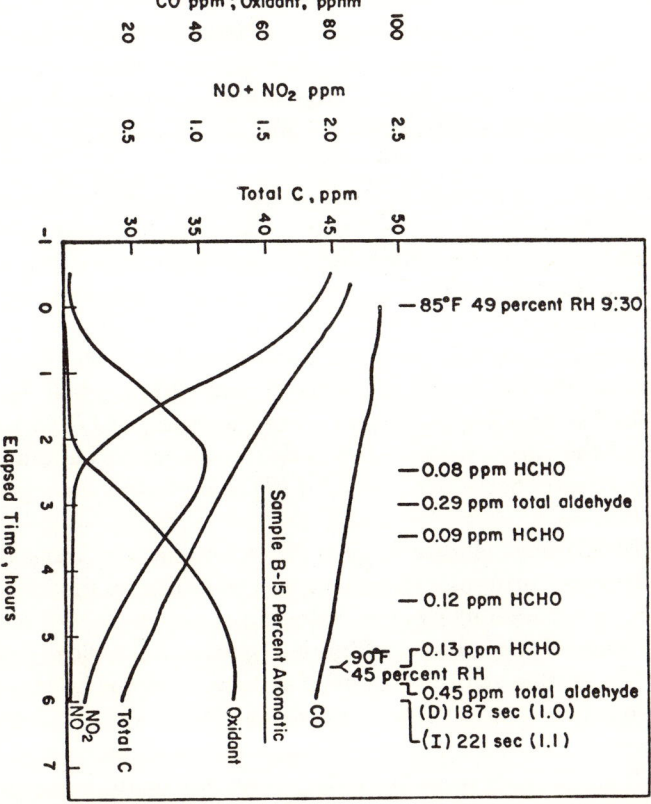

National Paint and Coatings Association

Figure 1. Typical smog profile—mineral spirits (4.0 ppm) + NO (2.0 ppm) (5)

Rule 66 provides a simple reactivity scale. It requires an 85% reduction of solvents emissions from various industrial application if these solvents are photochemically reactive. Photochemically reactive solvents contain:

(1) 5% or more olefinic hydrocarbons, alcohols, aldehydes, esters, ethers, or ketones,

(2) 8% or more of C_8 or higher aromatics, except ethylbenzene,

(3) 20% ethylbenzene, toluene, branched ketones, or trichloroethylene, or

(4) A total of more than 20% of the above.

When the Los Angeles studies were being conducted, the two principal criteria for defining photochemical reactivity were eye irritation and oxidant formation. By definition a photochemically reactive condition or photochemical smog condition existed if the average oxidant in Los Angeles were 0.1 ppm for 1 hour. Since then the National Air Quality Standards have reduced that level to 0.08 ppm ozone for 1 hour. The early Los Angeles work did attempt to establish general regions of reactivity for some solvents. However, since the Los Angeles group was a pioneer in this field, it recognized that there were unresolved technical problems in establishing reactivity data. Background air used in the Los Angeles studies for example was not clean and was difficult to clean. As a result, the background levels of eye irritation and of oxidant formation were high, and interpretations from these studies were complicated. In the present paper two major studies are reviewed. One study was conducted by Battelle-Columbus under the sponsorship of the National Paint, Varnish and Lacquer Association (5) (now the National Paint and Coatings Association), and the other was conducted at Stanford Research Institute under the auspices of the Environmental Protection Agency (6). For both studies, new smog chambers were constructed (7, 8) which took advantage of the earlier experiences of the Los Angeles Laboratories (9), the Public Health Service chamber (10), and other chambers in the country (11).

In most chamber studies 1–10 ppm of organic are allowed to react with 0.1–3.0 ppm nitric oxide. In this way a profile of the species being consumed or produced in the atmosphere can be developed, and various reactivities can be determined. Figure 1 presents a typical profile from this method. In this figure, 4 ppm of a mineral spirit containing 15% aromatics reacted with 2 ppm nitric oxide. NO_2 was produced as the hydrocarbon and nitric oxide was being consumed, and about the time that the NO was totally consumed, the ozone or oxidant began to appear. Two important items are the NO_2 maximum and the initial appearance of oxidant. Ozone makes up 80–90% of the total oxidant, and the two terms are often used interchangeably. All NO in the system is not ac-

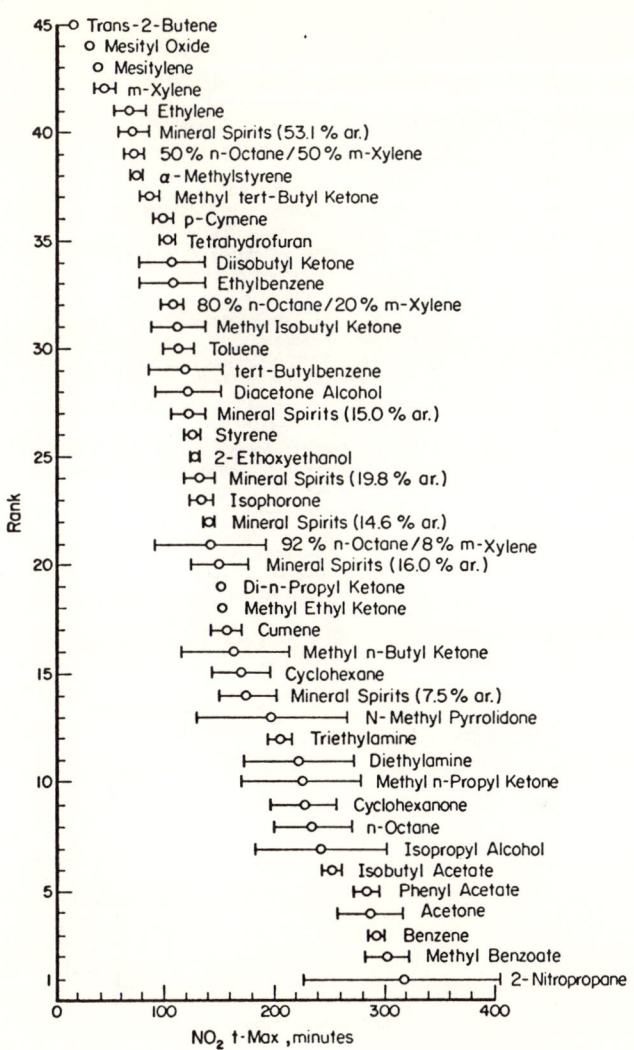

Figure 2. Ranking Based on NO_2 t-max (5)

National Paint and Coatings Association

counted for as NO_2 because numerous nitrates are also being produced during the reaction. The initial appearance of oxidant, coincident with the maximum NO_2, results from an important reaction in the smog process:

$$NO + O_3 = NO_2 + O_2$$

Because of this rapid and efficient reaction, ozone does not build up to any significant level until most of the NO is consumed.

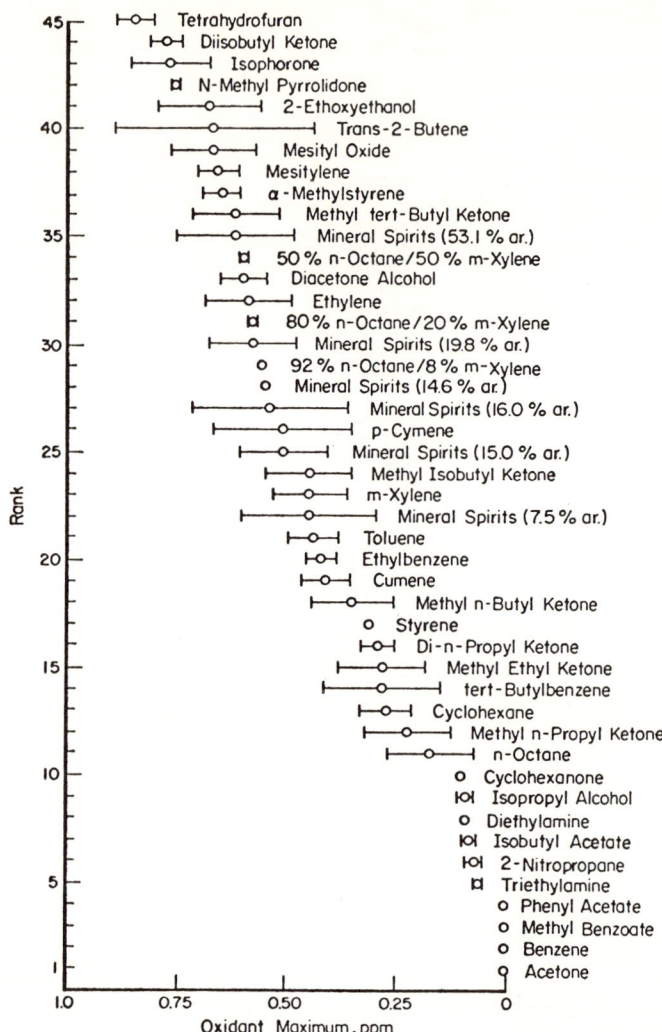

Figure 3. Ranking based on oxidant production (5)

At the top of Figure 1, various points in the reaction are noted where samples were removed for specific analysis—*i.e.*, for formaldehyde and total aldehyde. At the end of the reaction, one observes the eye-irritation response. Two panels of seven people each were used, and the average of each panel is shown at the top near the 6-hour reaction period.

The CO profile in Figure 1 is a measure of dilution in the chamber mix and is not a measure of reaction. CO can be considered nonreactive in these systems and can be used as a tracer gas to follow the replenishment of chamber air with clean air in batch-type chamber reactions. (Evidence that CO does participate in the reaction has been observed by several investigators (12, 13, 14), but that does not preclude the use of CO in the above role.)

Using this type of reaction profile, various methods have been used to define reactivity. In this paper, the discussion is limited to the rate of photooxidation of nitric oxide (expressed as the NO_2 t-max, the time to reach maximum NO_2), the maximum oxidant produced or the dosage of

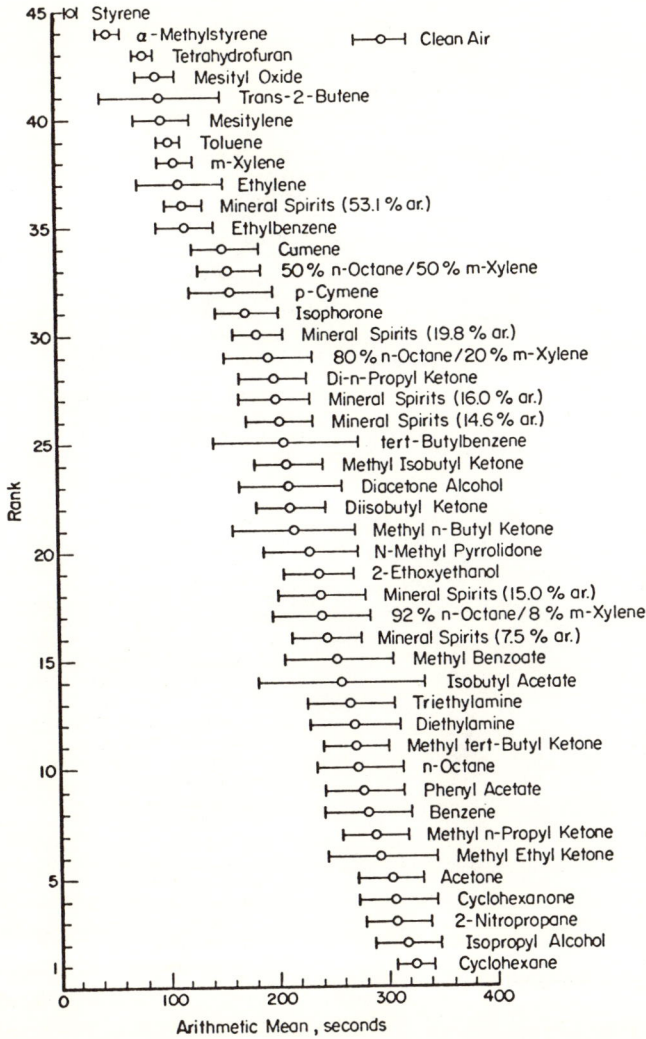

National Paint and Coatings Association

Figure 4. Ranking based on eye–response data (5)

oxidant (the ppm–time area under the oxidant curve), eye-response time, and formaldehyde formation.

Battelle Solvent Study

In the program conducted for the National Paint, Varnish and Lacquer Association, 45 solvents were examined under the following conditions: 4 ppm organic, 2 ppm nitric oxide. Profiles were obtained for the solvents, and reactivity measurements were obtained from the profiles.

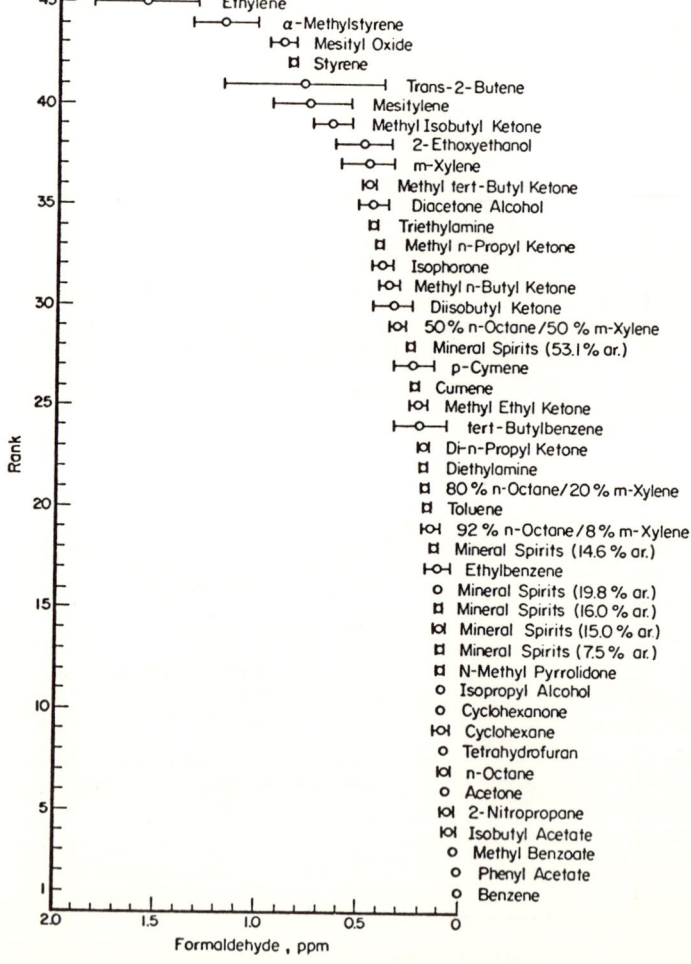

Figure 5. Ranking based on formaldehyde production (5)

National Paint and Coatings Association

Table I. Category Summary (5)

Solvents	NO_2 t-Max	Oxidant Max	Eye Response	Formaldehyde	Averaged Rating
Acetone	1	1	1	1	1
Benzene	1	1	1	1	1
trans-2-Butene	3	3	2	2	2.5
tert-Butylbenzene	2	1	1	1	1.2
Cumene	2	2	2	1	1.7
Cyclohexane	2	1	1	1	1.2
Cyclohexanone	1	1	1	1	1
p-Cymene	3	2	1	1	1.7
Diacetone alcohol	2	3	1	1	1.7
Diethylamine	1	1	1	1	1
Diisobutyl ketone	2	3	1	1	1.7
Di-n-propyl ketone	2	2	1	1	1.5
2-Ethoxyenthanol	2	3	1	1	1.7
Ethylbenzene	2	2	2	1	1.7
Ethylene	3	3	2	3	2.7
Isobutyl acetate	1	1	1	1	1
Isophorone	2	3	1	1	1.7
Isopropyl alcohol	1	1	1	1	1
Mesitylene	3	3	2	2	2.5
Mesityl oxide	3	3	2	2	2.5
Methyl benzoate	1	1	1	1	1
Methyl tert-butyl ketone	3	3	1	1	2
Methyl isobutyl ketone	2	2	1	2	1.7
Methyl n-butyl ketone	2	2	1	1	1.5
Methyl ethyl ketone	2	1	1	1	1.2
Methyl n-propyl ketone	1	1	1	1	1.0
N-Methyl pyrrolidone	2	3	1	1	1.7
α-Methylstyrene	3	3	3	3	3
Mineral spirits 7.5% ar.	2	2	1	1	1.5
14.6% ar.	2	2	1	1	1.5
15.0% ar.	2	2	1	1	1.5
16.0% ar.	2	2	1	1	1.5
19.8% ar.	2	3	1	1	1.7
53.1% ar.	3	3	2	1	2.2
2-Nitropropane	1	1	1	1	1
n-Octane	1	1	1	1	1
92% n-Octane/8% m-Xylene	2	2	1	1	1.5
80% n-Octane/20% m-Xylene	2	3	1	1	1.7
50% n-Octane/50% m-Xylene	3	3	2	1	2.2
Phenyl acetate	1	1	1	1	1
Styrene	2	2	3	2	2.2
Tetrahydrofuran	2	3	2	1	2
Toluene	2	2	2	1	1.7
Triethylamine	1	1	1	1	1
m-Xylene	3	2	2	1	2

The four reactivities discussed here are NO_2 t-max, maximum oxidant, eye response, and formaldehyde production.

Figure 2 shows the 45 solvents ranked with respect to NO_2 t-max. The solvents are ranked from highest reactivity to the lowest. Two olefinic materials are the most reactive (Figure 2) followed by two aromatics, mesitylene and m-xylene. Further down the list as reactivity decreases oxygenated materials appear at the lower end of the reactivity scale, and the ketones and esters show up as the least reactive in terms of NO_2 t-max.

Figure 3 ranks the solvents according to maximum oxidant formation. Tetrahydrofuran and diisobutyl ketone are the most reactive. *trans*-2-Butene and mesityl oxide, which rank highest in NO_2 t-max, are in positions 40 and 39, respectively in this oxidant scale. The ketones, esters, and alcohols rank lowest in oxidant production.

Figure 4 ranks eye-response as the arithmetic mean response time. Styrene and α-methylstyrene are the most reactive species partly because of their benzyl groups. Heuss and Glasson (15) showed that peroxybenzoylnitrate (PBzN), a specific lachrymator produced in these systems, is 200 times more reactive than peroxyacylnitrate, which itself is a fairly potent lachrymator. Tetrahydrofuran, mesityl oxide, and *trans*-2-butene are high in eye irritation (Figure 4).

Clean air is also ranked in Figure 4 at an average eye response of 300 sec. Six solvents are as irritating or less irritating than clean air. Eye response is very subjective, and in my opinion response times over 200 sec indicate a low irritation factor. Because of this subjectivity, these six solvents might be considered non-eye-irritating solvents.

Figure 5 shows ratings based on formaldehyde production. Romanovsky *et al.* (16) noted a significant correlation between eye-response time and formaldehyde production.

A comparison of Figure 5 with Figure 4 suggests that the scatter in eye irritation response indicates a weakening relationship between eye irritation and formaldehyde concentration. Tetrahydrofuran, toluene, and ethylbenzene exhibit high eye irritation and low formaldehyde concentra-

Table II. Category Ranges

Reactivity Category	NO_2 t-Max, min	Oxidant Max, ppm	E. I. sec[a]	Formaldehyde, ppm
3 high reactivity	0–100	0.57–0.85	0–60	1.0–1.6
2 intermediate reactivity	101–200	0.29–0.56	61–160	0.5–1.0
1 Low reactivity	201–360	0–0.28	161–360	0–0.5

[a] Eye response.

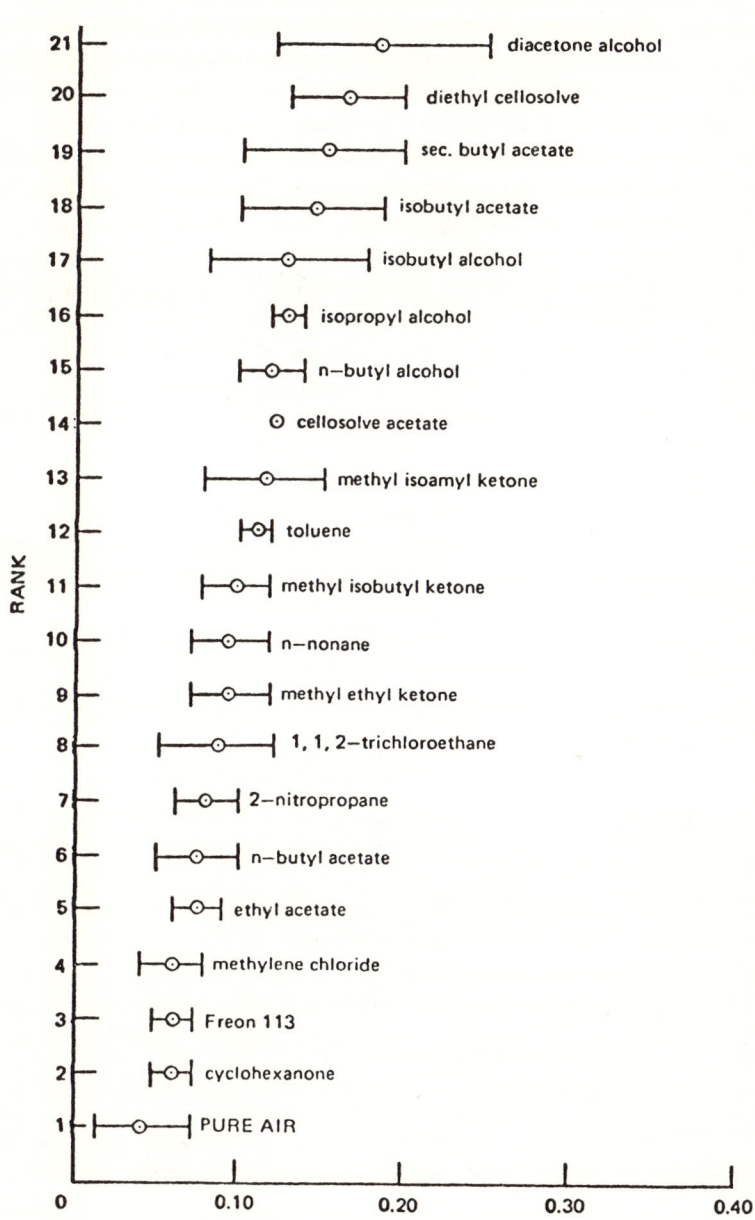

Figure 6. Rankings based on normalized maximum oxidant (6)

80   SOLVENTS THEORY AND PRACTICE

tion. Although toluene and ethylbenzene contain benzyl structures, other lachrymators such as PAN and PBzN also contribute to eye irritation (*15*).

Category Summary. Ranking individual solvents by parameter categories as in Figures 2–5 is not very practical. Table I evaluates reactivities of the solvents on a scale of 1-3; high, intermediate, or low reactivity (as defined by Table II). This provides a rather broad classification.

In Table I only a few solvents exhibit the same ranking on all four scales. Low-category solvents however are fairly consistent in their rankings. Acetone, benzene, cyclohexanone, diethylamine, isobutyl acetate,

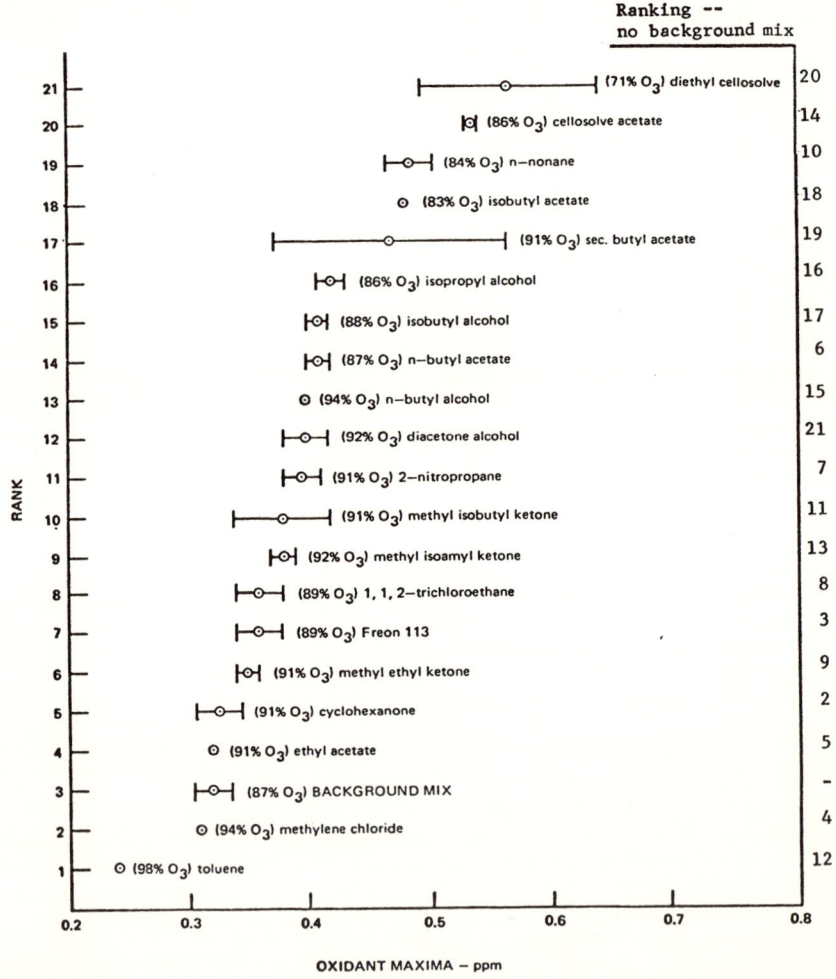

Figure 7. Rankings based on oxidant maxima in runs with background mix plus 8 ppm test solvent (6)

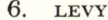

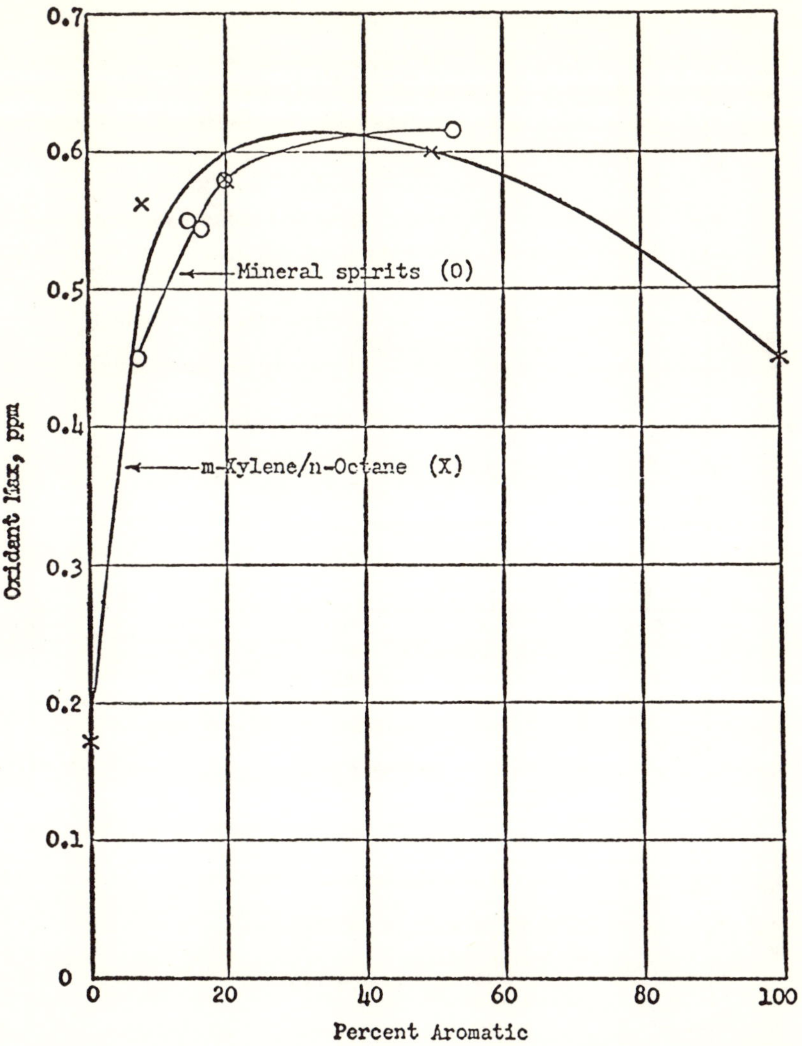

Environmental Science Technology

Figure 8. Oxidant max as a function of aromatic content of mixtures

methyl benzoate, 2-nitropropane, *n*-octane, phenyl acetate, and triethylamine exhibited low reactivity in all categories. α-Methylstyrene exhibited high reactivity in all categories. In most cases the specific reactivity depended on the type of reactivity response being examined. This illustrates the complexity of attempting to define solvent reactivity for any specific application. The chemical reactivity of benzene was low in these four categories examined, but in another recent study benzene was a significant contributor to atmospheric aerosol (*17*).

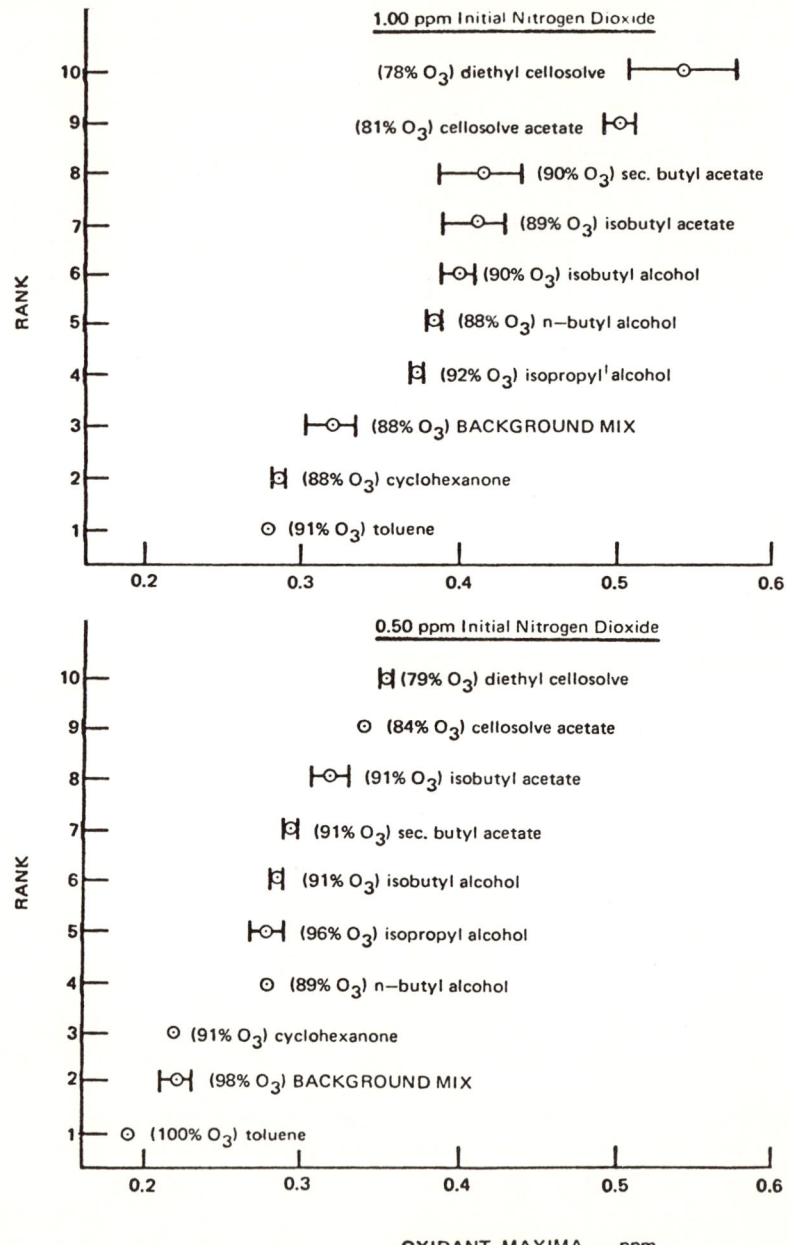

Environmental Science Technology

Figure 9. Rankings based on oxidant maxima in runs with background mixes plus 4 ppm test solvent (6)

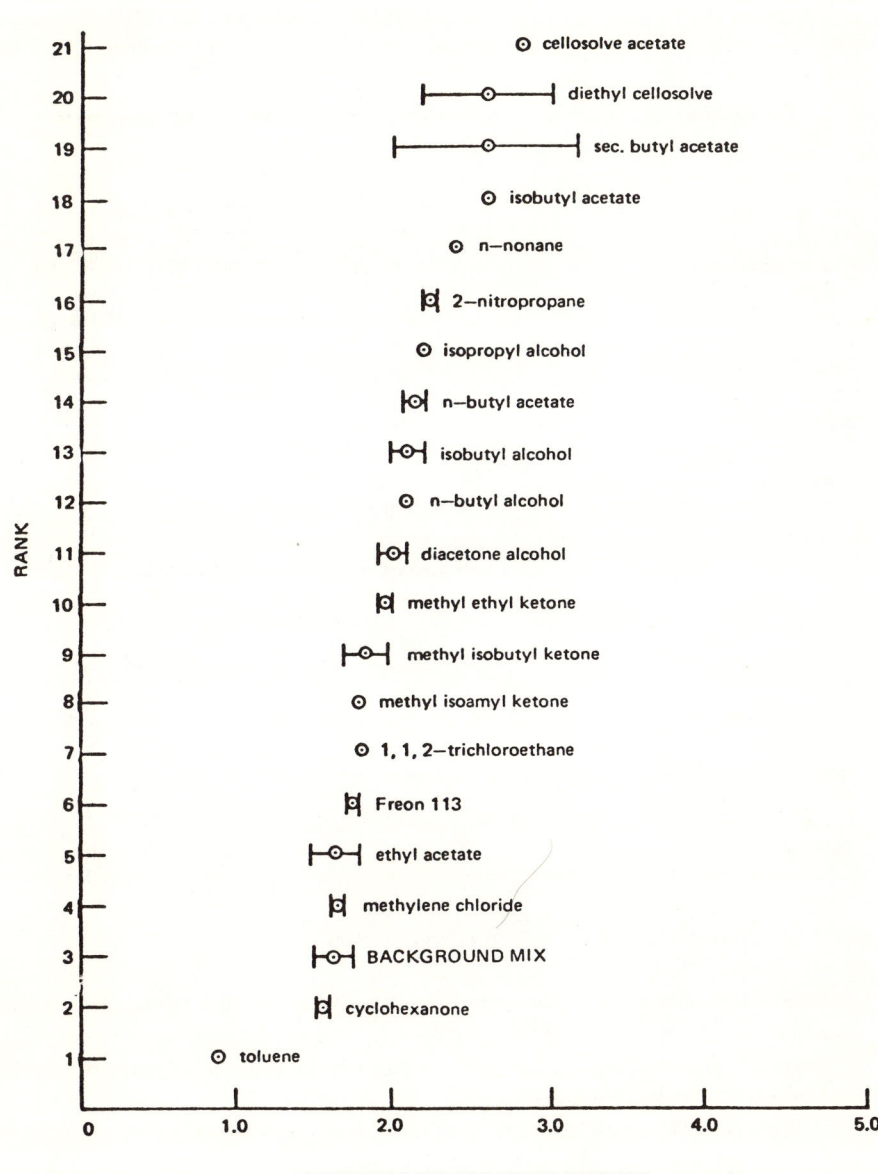

Environmental Science Technology

Figure 10. Rankings based on oxidant dosages in runs with background mix plus 8 ppm test solvent (6)

Stanford Research Institute Studies

The solvent program at SRI was carried out in ways similar to the Battelle-Columbus study. In the Stanford studies the solvents were examined with a clean-air background and with a hydrocarbon-mixture air

Table III. Comparison of Oxidant–Maximum Rankings, ppm

	BCL_s		SRI_n		SRI^c	
MIBK	0.45	3	0.13	4	0.34	6
MEK	0.28	5	0.13	5	0.35	5
Cyclohexanone	0.10	6	0.085	8	0.31	7
Diacetone alcohol	0.60	2	0.24	1	0.40	3,4
Isopropyl alcohol	0.09	7	0.12	6	0.40	3,4
Isobutyl acetate	0.08	8	0.15	3	0.44	2
Diethyl Cellosolve	0.68	1	0.22	2	0.55	1
Toluene	0.44	4	0.08	9	—	—
2 Nitropropane	0.07	9	0.10	7	—	—

[a] 4 ppm solvent/2 pp, NO.
[b] 1 ppm solvent/0.1 ppm NO_x.
[c] 8 or 4 ppm solvent/1 ppm NO_x, plus gackground mixture.

Table IV. Oxidant Dosage Rankings, ppm-min

	BCL^a		SRI^b	
MIBK	267	2	36	3
MEK	232	5	21	7,8,9
Cyclohexanone	262	2	21	7,8,9
Diacetone alcohol	193	7	62	2
Isopropyl alcohol	215	6	33	5
Isobutyl acetate	252	4	45	4
Diethyl Cellosolve	274	1	75	1
Toluene	141	9	30	6
2-Nitropropane	180	8	21	7,8,9

[a] 4 ppm solvent/2 ppm NO, 6 hr dosage.
[b] 1 ppm solvent/0.1 ppm NO_x, 7 hr dosage.

background. The conditions were 1 ppm solvent and 0.1 ppm nitric oxide. Ranking data for 20 solvents are given in Figure 6, from a No. 21 (high reactivity) to pure air. Diacetone alcohol is highest in reactivity; cyclohexanone is the lowest next to pure air.

For solvents studied with a hydrocarbon background mixture, two mixtures were used: propylene (0.5 ppm), n-butane (2 ppm), toluene (1 ppm) and NO_2 (1 or 0.5 ppm). Eight ppm of solvent were used in each experiment. Figure 7 tabulates the results. Also shown on this figure (on the right-hand side) is the relative ranking position that the solvent had when it was examined in a clean-air background (*see* Figure 6). A

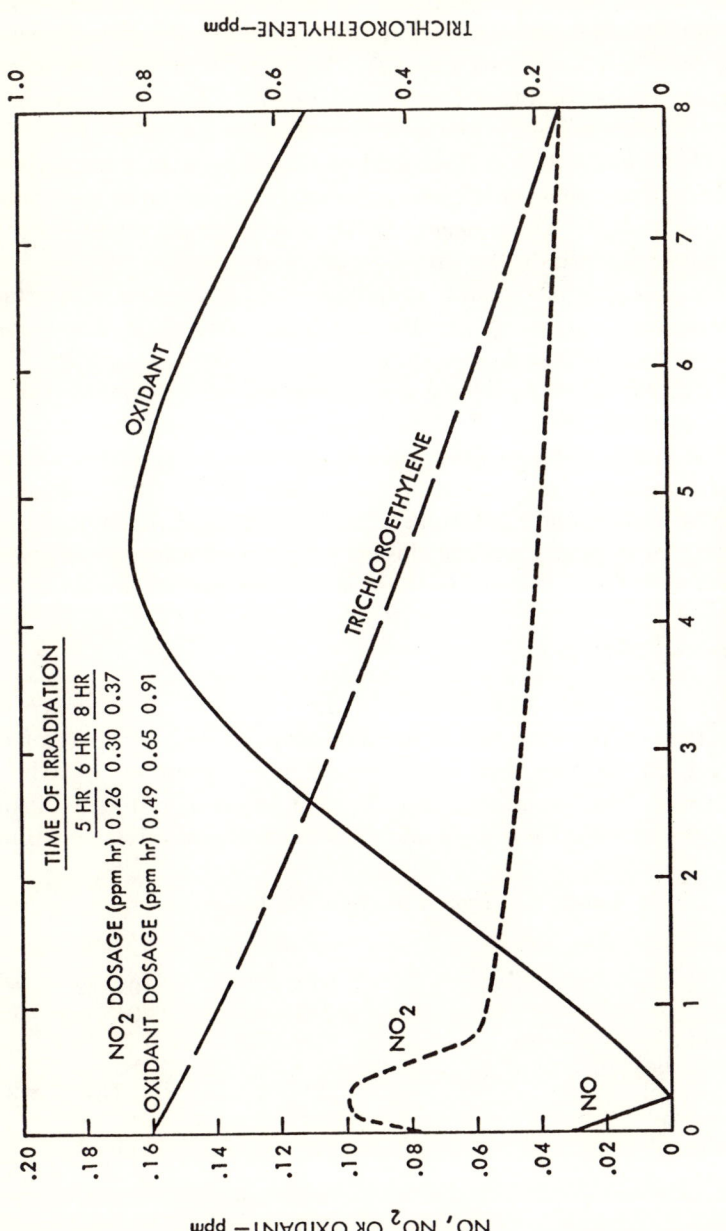

Figure 11. *Concentration changes during photooxidation of trichloroethylene (0.8 ppm) with nitrogen oxides (0.1 ppm) (19, 20)*

change of two or three positions is not significant. A change of five or six positions is. In Figure 7, eight materials were shifted markedly in going from a clean-air background to a specific background mixture (Cellosolve acetate, n-nonane, isobutyl alcohol, n-butyl acetate, diacetone alcohol, Freon 113, cyclohexanone, and toluene). The first three solvents moved up in ranking, while the other five moved down. Of special interest is n-nonane. Although it was No. 10 solvent in Figure 6, it is No. 19 in Figure 7, fairly high in reactivity. This shift in reactivity with a background mixture that contains an aromatic and an olefin is borne out by the studies at Battelle-Columbus. There seems to be a synergistic effect when a paraffinic material reacts in the presence of an aromatic.

Figure 8 is a plot of oxidant formation when a mixture of n-octane and m-xylene react in smog. As the aromatic content of the system increases, the oxidant reaches a maximum near 30% aromatic. This synergistic effect is also supported by SRI's studies with n-nonane using a background mixture. Figure 8 also shows that esters tend to produce more oxidant while ketones produce less oxidant when either solvent reacts in the presence of background mixture. While the addition of toluene to the background mixture resulted in decreased reactivity (measured by oxidant max and oxidant dosage) there was a large increase in aerosol formation (as indicated by light scattering measurements).

Effect of Initial Nitrogen Dioxide

The quantity of nitrogen dioxide in the reaction will influence the overall reaction of any material in the atmosphere (*18*). As part of the SRI program, 10 solvents that were studied in the presence of a background mix were also examined with different levels of nitrogen dioxide (Figure 9); 4 ppm of solvent were added to background mixes containing

Table V. Eye-Irritation Rankings[a]

	BCL[b]		SRI[c]	
MIBK	213	2	162	2
MEK	295	6	201	5
Cyclohexanone	309	7,8	193	4
Diacetone alcohol	214	3	188	3
Isopropyl alcohol	318	9	<240	7,8,9
Isobutyl acetate	258	5	210	6
Diethyl Cellosolve	239	4	144	1
Toluene	106	1	<240	7,8,9
2-Nitropropane	309	7,8	<240	7,8,9
Air	280		<240	

[a] Response time, sec.
[b] 4 ppm solvent/2 ppm NO.
[c] 1 ppm solvent/0.1 ppm NO_x.

either 1 ppm or 0.5 ppm nitrogen dioxide. Results showed that although an increase in nitrogen dioxide resulted in an increase in the maximum oxidant produced, no real change in the ranking of these solvents resulted. Solvents behave essentially the same regardless of the initial NO_2 present.

Oxidant production can also be examined on the basis of oxidant dosage, namely the concentration–time area under the oxidant curve (Figure 10). In the SRI studies, there was essentially no difference in ranking based on the dosage relative to the oxidant max on the background mixtures. Two possible exceptions are 2-nitropropane and methyl ethyl ketone which shifted four or five positions. In general, however, one does not expect a shift unless there is a secondary reaction which rapidly depletes the oxidant.

Comparison of Solvent Data

As I have already said, there are a number of difficulties in reproducing data within specific smog chambers, especially in attempting to compare results from different chambers. A program is being conducted by Lockheed Missiles and Space Co., sponsored by EPA and the Coordinating Research Council, aimed at determining why experiments in different smog chambers often produce different results even though the experiments are conducted in supposedly the same manner. In the two studies discussed here, however, none of the solvents examined by both laboratories lend themselves to comparison, and although they are compared on a different basis with respect to the background air and concentrations used, the results are of some interest.

In Table III, the oxidant maxima for the Battelle-Columbus and SRI data are compared. Here and in the subsequent table, the specific data are shown in each column with the rankings of the data among the nine materials presented alongside the concentration or eye-irritation data. These rank from No. 1 to No. 9. (Where more than one solvent had the same ranking, multiple ranks are used). Oxidant maximum data for diethyl Cellosolve and diacetone alcohol agree for both laboratories. Likewise methyl isobutyl ketone maintains the same ranking. For toluene the Battelle data show a significantly higher oxidant production than the SRI data. Although the concentration of solvent used in the SRI study was much less than that used in the Battelle study, isopropyl alcohol, isobutyl acetate, and 2-nitropropane showed greater oxidant yields in the SRI study. Thus, we see that both the observed relative reactivity and absolute reactivity can be significantly affected by the solvent/NO_x ratio used in a reactivity study.

Table IV compares oxidant dosage of the two studies. Expected trends between oxidant maximum and oxidant dosage are not seen. Both

Table VI. SO_2–Solvent

Solvent, 4 ppm NO, 2 ppm

Solvent	SO_2, Conc. [b]	Eye Irritation [a]		
		Equiv. sec [c]	Severity Index [d]	Subjective Severity [e]
Clean Air	0	292	1.9	Clean Air
	1.6	298	1.7	Clean Air
Toluene	0	89	7.5	Moderate
	1.3	96	7.3	Moderate
Mineral Spirits	0	245	3.2	Slight
(7.5% aromatics)	1.5	196	4.6	Slight
Isophorone	0	156	5.7	Moderate
(cyclic ketone)	1.7	151	5.8	Moderate
Mesityl oxide	0	86	7.6	Moderate
(unsaturated ketone)	1.8	82	7.7	Moderate
Methyl isobutyl ketone	0	134	6.3	Moderate
	1.8	216	4.0	Slight

[a] Measured by panels of 7 or 14 people after 6 hours of irradiation unless indicated otherwise.
[b] Concentration in ppm (v/v).
[c] Equivalent seconds—antilog of the average of the logs of the individual times.

laboratories find diethyl Cellosolve the most reactive of the nine solvents and methyl isobutyl ketone either second or third in ranking. However, SRI observed diacetone alcohol as second in ranking for dosage, and Battelle observed a relatively low dosage, indicating that the oxidant produced is being consumed by other reactions.

Table V compares the two laboratory investigations in terms of eye irritation. Here significant differences are observed. Toluene yields the highest eye-irritation reactivity in the Battelle chamber and the lowest in the SRI chamber. For MIBK and diacetone alcohol, both laboratories agree. The other rankings are not significant since the eye-irritation responses are so close to those obtained for air.

Trichloroethylene

Trichloroethylene received special attention in the establishment of Rule 66. It did not contribute significantly toward eye irritation in the early Los Angeles studies; however there was evidence that it did contribute substantially toward oxidant formation. Therefore, since large quantities of trichloroethylene were being used, trichloroethylene was included in Rule 66.

In an attempt to clarify this problem the Manufacturing Chemists Association sponsored a study at SRI to establish the relative reactivity of trichloroethylene (19, 20). Figure 11 presents a profile for trichloro-

Experiments

Aerosol, log units, Sinclair-Phoenix	Oxidant, Max. Conc. [b]
0.1	—
2.6	—
3.0	0.35
3.1	0.36
1.4	0.40
2.8	0.43
3.2	0.71
3.1	0.63
0.1	0.78
3.1	0.66
0.1	0.43
1.1	0.36

[d] Based on [(360—response time)/360] × 10.
[e] Average panel response from the four categories: severe, moderate, slight, equivalent to clear air.

ethylene in a smog system containing 0.8 ppm trichloroethylene and 0.1 ppm nitric oxide. A significant quantity of ozone was produced (about 0.17 ppm), but the SRI investigators felt that this was not an especially significant amount of ozone (21). This yield is comparable with the most reactive solvents appearing in Figure 6.

Other Atmospheric Influences

Although most reactivity studies have been conducted with a controlled background air or a clean background air, other parameters, especially sulfur dioxide and relative humidity in the atmosphere, influence the atmospheric process. In Table VI and Figures 12, 13, and 14 the influence of SO_2 is shown on some solvent systems (5, 22, 23). In Table VI the eye irritation, with and without SO_2, is shown for five solvents; one cannot state *a priori* that SO_2 reduces or increases the eye-irritation factor. Adding SO_2 to the mineral spirit systems increased eye irritation whereas adding SO_2 to methyl isobutyl ketone reduced irritation. For the other three solvents shown, there is no significant difference between the clean air or the SO_2 clean air system.

For 1-butene the influence of relative humidity (RH) on SO_2 (Figure 12) is strong with respect to oxidant formation (22). In dry or 22% RH systems, an increase in SO_2 results in no change or only a slight increase in oxidant, whereas in a moist system the increase in SO_2 results in a

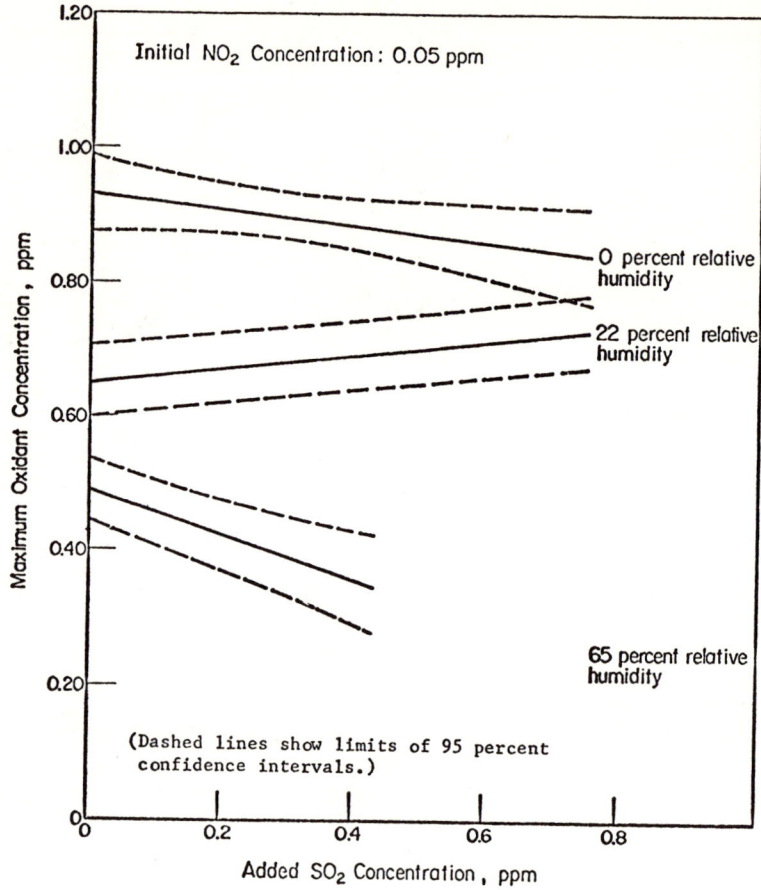

FIGURE 12. EFFECT ON OXIDANT FORMATION OF SO_2 ADDITIONS TO THE 1-BUTENE SYSTEM (Ref. 15)

4 ppm 1-butene
1 ppm NO

Journal of the Air Pollution Control Association

Figure 12. Effect on oxidant formation of SO_2 additions to the 1-butene system—4 ppm 1-butene, 1 ppm NO (18)

decrease in oxidant formation. Figure 13 however, shows an increase in oxidant formation in the toluene systems with the addition of SO_2. The effects of SO_2 on the smog system remain too complex for extensive interpretation. It is not possible, therefore, to incorporate the results of SO_2 additions from Table VI (where NO was 2 ppm and the reaction was conducted in a closed system) into the results of Figures 12 and 13 (where NO was 1 ppm and 0.5 ppm, respectively, and the reaction was conducted dynamically).

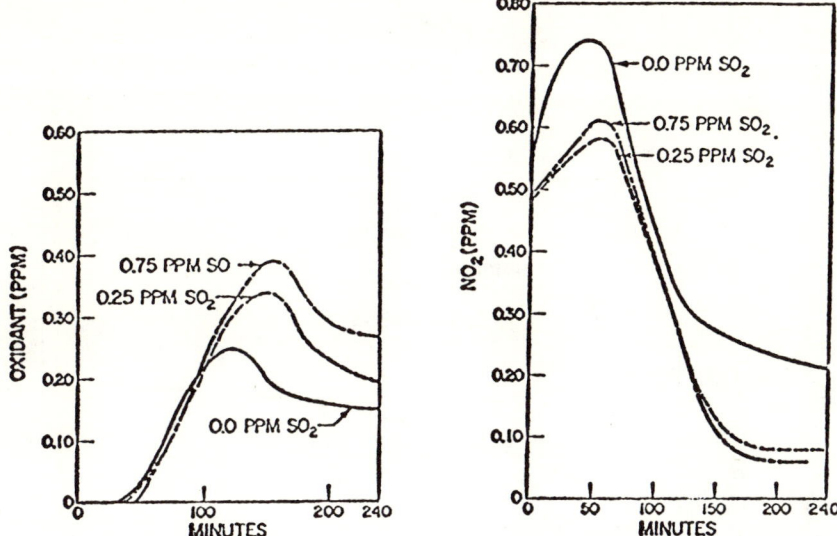

Figure 13. Effect of SO_2 on oxidant and NO_2 in the toluene–NO_x system— 4 ppm toluene, 0.5 ppm NO, 0.5 ppm NO_2, 50% RH (23)

Aerosols are becoming a greater concern in interpreting our atmospheric problems. Figure 14 is a compilation of some aerosol data with respect to SO_2 formation for hydrocarbon and solvent materials. Except for 1-heptene where aerosol formation increases drastically with only slight addition of SO_2, the aerosol production increases only slightly as SO_2 is increased in the system.

Solvent Reformulation

One of the major purposes, for examining the reactivities of solvents is to provide a basis for reformulating solvents that are less reactive and less polluting. Table VII presents a possible reformulation for reducing the reactivity of a solvent mixture. This reformulation is one of almost infinite possibilities. If the solvent to be reformulated is being used in a paint application, for example, many other requirements must be met. For the example in Table VII it was necessary to reduce the toluene level drastically to arrive at a lower reactivity, and in doing so, replacing 2-nitropropane with butylacetate raised the cost of the solvent significantly.

In reformulating a solvent mixture for lower reactivity, one must be careful that while complying with the law, he is not actually releasing more pounds of reactive solvent. Table VIII illustrates this situation.

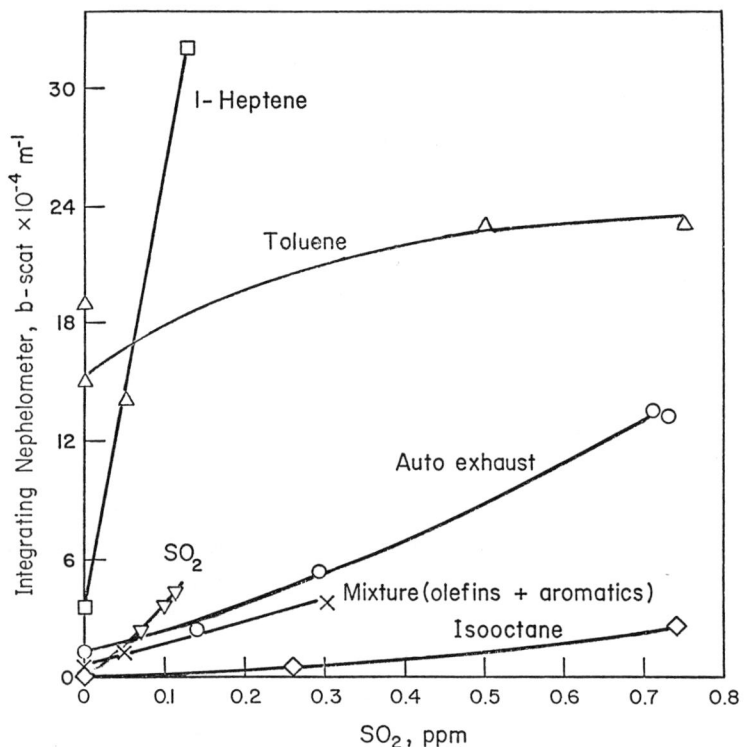

Figure 14. Influence of SO_2 on photochemical aerosol formation

Table VII. **Cost and Effectiveness of a Substitution for a Solvent Blend (24)**

Solvent		Reformulation	
Methyl ethyl ketone	14%	Methyl ethyl ketone	68%
Toluene	80%	Toluene	10%
2-Nitropropane	6%	Butyl acetate	22%
Cost	$0.52/gal	Cost	$0.94/gal
Reactivity	0.34 [a]	Reactivity	0.13 [a]

[a] Weighted reactivity by proportions of the compounds in the formula.

Table VIII. A Hypothetical Solids-to-Solvent Rule 66 Situation (25)

Coating	Ratio Between lbs Solvent-Solids	Ratio Between lbs Reactive Solvent-Solids
A Original paint 40% solids 60% solvent (35% of solvent reactive)	1.5 lbs solvent to 1.0 lbs solids	0.52 lbs reactive solvent to 1.0 lbs solids
B Reformulated paint 25% solids 75% solvent (20% of solvent reactive)	3.0 lbs solvent to 1.0 lbs solids	0.60 lbs reactive solvent to 1.0 lbs solids

Control of organic solvents emissions requires many considerations if we are to satisfy the demands of a clean environment and of high product quality.

Acknowledgment

The author gratefully acknowledges the comments and suggestion of Stanley L. Kopczynski, especially concerning the SRI solvent study in which he was Project Officer.

Literature Cited

1. Levy, A., Miller, S. E., "Problems in Defining Smog Reactivity," *Annu. Meetg. Air Pollut. Contr. Ass., 64th, Atlantic City, 1971,* paper 71-38.
2. Altshuller, A. P., Bufalini, J. J., "Photochemical Aspects of Air Pollution: A Review," *Environ. Sci. Technol.* (1971) **5,** 39.
3. Glasson, W. A., Tuesday, C. S., "Hydrocarbon Reactivities in the Atmospheric Photooxidation of Nitric Oxide," *Environ. Sci. Technol.* (1970) **4,** 917.
4. Brunelle, M. F., Dickinson, J. E., Hamming, W. J., "Effectiveness of Organic Solvents in Photochemical Smog Formation," Los Angeles Air Pollution Control District, July 1966.
5. Levy, A., Miller, S. E., "The Role of Solvents in Photochemical Smog Formation," Nat. Paint Varn., Lacquer Ass., Circ. 799 (April 1970).
6. Wilson, K. W., Doyle, G. J., "Investigation of Organic Solvents," Final Report, EPA Contract CPA 22-69-125 (September 1970).
7. Scofield, F., Levy, A., Miller, S. E., "Design and Validation of a Smog Chamber," Nat. Paint, Varn., Lacquer Ass., Circ. 797 (January 1969).
8. Doyle, G. J., "Design of A Facility (Smog Chamber) for Studying Photochemical Reacticns under Simulated Tropospheric Conditions," *Environ. Sci. Technol.* (1970) **4,** 916.

9. Griswold, S. S., Atkisson, A. A., Neligan, R. E., Bonamassa, F., Linville, W., "Research and Test Facilities of the Los Angeles County Air Pollution Control District," *J. Air Pollut. Contr. Ass.* (1963) **13**, 112.
10. Rose, A. H., Jr., Brandt, C. S., "Environmental Irradiation Test Facility," *J. Air Pollut. Contr. Ass.* (1960) **10**, 331.
11. Tuesday, C. S., D'Alleva, B. A., Heuss, J. M., Nebel, G. J., "The Smog Chamber," *General Motors Res. Rep.* **490** (June 1965).
12. Stedman, D. H., Morris, E. D., Jr., Daby, E. E., Niki, H., Weinstock, B., "Abstracts of Papers," 160th Meeting, ACS, Sept. 1970, WATR 26.
13. Westberg, K., Wilson, K. W., "Carbon Monoxide: Its Role in Photochemical Smog Formation," *Science* (1971) **171**, 1013.
14. Wilson, W. E., Jr., Ward, G. F., "Abstracts of Papers," 160th National Meeting, ACS, Sept. 1970, WATR 28.
15. Heuss, J. M., Glasson, W. A., *Environ. Sci. Technol.* (1968) **2**, 1109.
16. Romanovsky, J. C., Ingels, R. M., Gordon, R. J., *J. Air Pollut. Contr. Ass.* (1967) **17**, 454.
17. Wilson, W. E., Jr., Miller, D. F., Levy, A., "A Study of Motor Fuel Composition Effects on Aerosol Formation," Summary Report, Amer. Petrol. Inst. (Publication 4147), Feb. 21, 1972.
18. Wilson, W. E., Jr., Levy, A., *J. Air Pollut. Contr. Ass.* (1970) **20**, 385.
19. Wilson, K. W., "Photoreactivity of Trichloroethylene," Summary Report, Mfg. Chem. Ass., Sept. 1969.
20. Wilson, K. W., *J. Air Pollut. Contr. Ass.* (1969) **19**, 968.
21. Chass, R. L., Letter to the Editor, *J. Air Pollut. Contr. Ass.* (1970) **20**, 104.
22. Wilson, W. E., Jr., Levy, A., McDonald, E. H., *Environ. Sci. Technol.* (1972) **6**, 423.
23. Wilson, W. E., Jr., Levy, A., Wimmer, D. B., *J. Air Pollut. Contr. Ass.* (1972) **22**, 27.
24. Gamble, B. L., "Control of Organic Solvent Emissions in Industry," Ann. Meet., APCA, 1st, St. Paul, 1968.
25. Zimmt, W., E. I. DuPont Co., Marshall Laboratories, Philadelphia, private communication.

RECEIVED May 5, 1972.

Photochemical Smog and the Atmospheric Reactions of Solvents

JOHN L. LAITY,[1] ISRAEL G. BURSTAIN, and BRUCE R. APPEL

Shell Development Co., Emeryville, Calif. 94608

This paper presents the chemical measurements observed in our smog chambers for individual irradiations of solvent components. To correlate chemical reactivity with structure, we review many observations from other investigations of solvent reactivities, and we present generalizations about the smog-forming tendencies of solvent ingredients. Olefins and aromatics decompose primarily from electrophilic attack at unsaturated sites, and chemical reactivity therefore is raised by electron-donating groups on the double bond or ring. Alcohols, ethers, and esters react at an α-carbon atom, and reactivity is reduced if no hydrogens are present on the α-carbon. With ketones, chemical reactivity apparently is increased by the presence of alkyl or other stabilizing groups on the carbon atoms adjacent to the carbonyl.

The advent of Los Angeles County's Rule 66, as well as similar regulations which attempt to limit the amounts and compositions of solvents emitted into the atmosphere, has focused considerable attention on the smog-forming tendencies of solvents. A primary goal of smog research in several laboratories has been to develop a scale of atmospheric reactivities of solvent components. Hopefully such a scale, when coupled with an evaluation of air pollution sources, would allow at least qualitative estimation of the amount and type of material that could be released to an atmosphere without resulting in intolerable levels of photochemical smog.

[1] Present address: Shell Development Co., MTM Product Research and Development Laboratory, P.O. Box 262, Wood River, Ill. 62025.

In areas suffering from photochemical smog, the atmospheric reactions and effects of solvents are invariably dwarfed by reactions and effects of automotive emissions. An actual urban atmosphere is simply too unwieldy for controlled, scientific studies of many solvent reactions. Hence, laboratory irradiation chambers are commonly used to assess solvent reactivities under simulated conditions. Chemical reactions of photochemical smog formation are unquestionably influenced by experimental conditions and constraints of each irradiation chamber, thus greatly complicating attempts to assign absolute reactivities to solvents. However, from studies of many solvent materials under fixed conditions (varying only the solvent component under study), it is possible to derive the relative contributions of solvent ingredients to several aspects of photochemical smog production.

Although concern about automotive emissions has led to a large number of useful reports [see, for example, Glasson and Tuesday (1)] on the reactivities of hydrocarbons in irradiation chambers, there are few comprehensive publications devoted to the contributions that solvents can make to photochemical smog. Los Angeles County's Air Pollution Control District (LA-APCD) evaluated several solvents (2, 3) in a smog chamber in the early 1960's, and Battelle Memorial Institute, in work sponsored by the National Paint, Varnish and Lacquer Association, recently investigated a variety of solvents and solvent components (4). The LA-APCD study concentrated mainly on the contributions of solvents to eye irritation and ozone formation in photochemical smog, while Battelle examined eye irritation and several chemical manifestations of smog.

Our solvent studies, which began in 1967, have involved only chemical measurements which are related to the production of photochemical smog. The most obvious symptoms of photochemical smog are eye irritation, visibility reduction, and plant damage, but the connections between such physical or biological smog symptoms and chemical measurements (like the rate of nitric oxide photooxidation) are tenuous (5). Results presented here take on additional significance since the observation by several groups (6, 7, 8) that carbon monoxide can be mildly reactive in smog formation. Several investigators (4, 9, 10) have used relatively high concentrations of carbon monoxide in their irradiation chambers as a measure of air replenishment. Our chamber studies were conducted in the absence (<1 ppm) of carbon monoxide.

This paper presents some of the chemical measurements observed in our chambers in 200 experiments involving approximately 50 individual compounds, most of which are commercial solvent components. To correlate chemical reactivity with structure, we also review many of the observations and conclusions of other investigations into solvent reactivities. Although the chemical reactions leading to photochemical smog are very complex and not completely understood, the principles presented here account for the reactivity differences observed among classes of solvent components including hydrocarbons, alcohols, ethers, esters, amides, and ketones.

Experimental

Irradiation Chambers. Most of the data were determined in a 397-liter vessel constructed chiefly of stainless steel. This chamber is a polished stainless steel cylinder 4-feet long and 2-feet in diameter with blacklight fluorescent lamps inside. Temperature is controlled by circulating liquid in coils around the chamber from a constant temperature bath. Several ports are on the chamber for removing and analyzing air samples. The general cleaning procedure between experiments was to evacuate the vessel to 1μ Hg and heat overnight at 50°–65°C. Values for the half-life of ozone (1 ppm) in the chamber air varied from 1.5 to 6 hours in the dark and 1.3 to 2.5 hours with the lights on. The rate of thermal oxidation of nitric oxide (1.5 ppm) in the chamber air is 1.7 (± 0.3) $\times$ 10^4 liter^2 mole^{-2} sec^{-1}. Light intensity is measured by the first-order rate constant for photolytic decomposition of nitrogen dioxide, giving $K_d = 0.40$ min^{-1} during the first 2 minutes of irradiation. The chamber and its characteristics were described previously (11). Two glass irradiation chambers (a 235-liter vessel and a 23-liter flask) used in our investigations are also described elsewhere (11, 12).

During dosing, samples of solvent and doubly-distilled water are vaporized into the evacuated chamber through a glass vacuum manifold. Nitric oxide gas in nitrogen (Matheson) is passed in, and Ultrapure air (Air Products and Chemicals) is then used to purge the manifold and fill the chamber to atmospheric pressure. The pure air supply contains <2 ppm methane as the only hydrocarbon, <1 ppm carbon monoxide, and <0.01 ppm nitric oxide or nitrogen dioxide; the hydrocarbon and nitrogen oxides content of each air cylinder was checked and found to be within these limits. Little or no oxidant (<0.01 ppm maximum) is produced in any of the chambers by irradiating the pure air supply for 5 hours.

Chamber Conditions. All experiments are conducted with stirring and irradiation for 5 hours. The chamber temperature before irradiation is 22°C; during irradiation the temperature of the steel chamber is maintained at 32°C, and the contents of the glass chambers are held at 27°C. All experiments begin with 1.5 ppm (by volume) solvent and 0.6 ppm nitrogen oxides (0.57 ppm nitric oxide and 0.03 ± 0.02 ppm nitrogen dioxide). Increases in relative humidity promote the reactions observed in the chambers (11), and the relative humidity is therefore regulated to 20% in the steel chamber and 60% in the glass vessels. Atmospheric pressure is maintained in each chamber during an experiment by using pure air to replace each sample withdrawn for analysis. The resulting dilution (never more than 10% in the larger chambers) is corrected in all calculations.

Analytical Techniques. Hydrocarbons are analyzed by GLC with a flame ionization detector calibrated with hexane standards of known ppm concentrations. Alkyl nitrates and peroxyacyl nitrates are monitored by GLC at room temperature on a 3-foot, 2-mm id glass column containing 5% Carbowax 600 on Chromosorb W, AW/DMCS. Peroxybenzoyl nitrate is separated on a 50-cm, 3-mm id glass column with 3% JXR on Gas Chrom Q. Quantitative analysis of peroxyacyl nitrates is achieved with an electron capture detector through infrared calibration and dilution of the known mixtures to the 10^1–10^2 ppb level.

Presence of nitrogen dioxide is determined colorimetrically with Saltzman reagent. Grab bottles (1 or 2 liters) are used to remove samples from the stainless steel chamber; the bottles are injected with 10 ml of reagent, allowed to develop for about 1 hour, and analyzed in 2.5 cm cells. For the small glass chamber the NO_2 analysis is conducted in miniature using 250 or 500 ml grab bottles, 6 ml of reagent, and a 10 cm cell. Nitric oxide is oxidized to NO_2 over dichromate-impregnated glass fiber paper before analysis with Saltzman reagent.

Formaldehyde levels are determined by the chromotropic acid method using midget impingers to collect samples. Total oxidant is measured periodically with a coulometric ozone meter (Mast Development Co.). The response of the meter to ozone, nitrogen dioxide, and peroxyacyl nitrate is determined by calibration experiments, and the oxidant level is expressed as ozone, with the contribution of nitrogen dioxide removed.

Treatment of Data. Chemical measurement data in each irradiation chamber experiment are plotted as concentration *vs.* time, and the rates and dosages are determined from the points and best-fit experimental curves. The name of the compound and initial chamber concentrations are entered on computer cards along with the following observations:

(1) maximum rate of NO_2 formation, (2) average rate of NO_2 formation, (3) time required to obtain one-half the NO_2 formed, (4) time to NO_2 maximum, (5) maximum NO_2, (6) maximum rate of NO_2 disappearance, (7) NO_2 dosage, (8) maximum rate of hydrocarbon disappearance, (9) average rate of hydrocarbon disappearance, (10) per cent hydrocarbon consumed in run, (11) time for consumption of 25% initial hydrocarbon, (12) maximum formaldehyde, (13) formaldehyde yield from hydrocarbon, (14) maximum oxidant, (15) time to maximum oxidant, (16) oxidant dosage, and (17) maximum peroxyacetyl nitrate. Since chamber performance is variable, we ran toluene frequently as a check of chamber performance. The appropriate toluene standard was noted for each run, and each chemical measurement is expressed relative to toluene or in "toluene equivalents." Since the results are available on computer cards, different techniques (*e.g.*, correlation analyses, rank ordering, etc.) for data analysis can be applied.

Results and Discussion

Results of Investigations in the Steel Chamber. Because of changes in chamber cleanliness, surface effects, chamber dosing procedures, irradiation source intensity, and other chamber parameters, results have varied in the steel chamber (*11*). The results given in Tables I to III are expressed relative to toluene (*e.g.*, maximum oxidant for the solvent

Table I. Relative Rates of Nitrogen Dioxide Formation in the Steel Chamber

Compound	Maximun Rate of NO_2 Formation (relative to toluene)
Trifluorotoluene	0.20
Benzene	0.20
tert-Butyl acetate	0.25
tert-Butyl alcohol	0.25
Acetone	0.3
N,N-Dimethylformamide	0.4
tert-Butylbenzene	0.4
tert-Amylbenzene	0.45
Isopropyl alcohol	0.45
Ethyl acetate	0.5
Ethyl alcohol	0.5
n-Butylcyclohexane	0.55
Methyl ethyl ketone	0.55
n-Pentane	0.55
p-Menthane	0.6
n-Octane	0.6
Diethyl ketone	0.7
n-Butyl acetate	0.7
Cyclopentanone	0.7
Isooctane	0.7
Isopropyl acetate	0.75
Cyclohexanone	0.8
n-Butylbenzene	0.8
Ethylbenzene	0.8
Isobutylacetate	0.9
Toluene	1.00
n-Butyl alcohol	1.0
sec-Butyl alcohol	1.0
Isobutyl alcohol	1.1
Methyl n-butyl ketone	1.15
1,4-Diethylbenzene	1.2
4-Methyl-2-pentanol	1.4
Diethyl ether	1.45
Methyl isobutyl ketone	1.65
N,N-Dimethylacetamide	1.7
o-Xylene	1.80
Propylene	2.0
Tetrahydrofuran	2.3
m-Xylene	2.8
Mesitylene	3.0
2-Methyl-2-butene	7.8

Table II. **Relative Rates of Hydrocarbon Disappearance in the Steel Chamber**

Compound	Maximum Rate of Hydrocarbon Disappearance (relative to toluene)
Trifluorotoluene	0.05
tert-Butyl acetate	0.25
tert-Butyl alcohol	0.25
Acetone	0.25
Benzene	0.3
Ethyl acetate	0.3
Methyl ethyl ketone	0.3
Cyclopentanone	0.3
Ethyl alcohol	0.35
n-Butane	0.35
n-Pentane	0.35
Isopropyl acetate	0.4
Diethyl ketone	0.4
Octane	0.45
Isooctane	0.45
p-Menthane	0.5
Cyclohexanone	0.5
n-Butyl acetate	0.55
n-Butylcyclohexane	0.6
Isobutyl acetate	0.65
tert-Butylbenzene	0.7
Methyl n-butyl ketone	0.75
n-Butylbenzene	0.8
Ethylbenzene	0.8
Methyl isobutyl ketone	0.8
Isopropyl alcohol	0.85
Indane	0.85
Tetrahydrofuran	0.9
tert-Amylbenzene	0.9
Toluene	1.00
n-Butyl alcohol	1.2
4-Methyl-2-pentanol	1.2
Diethyl ether	1.5
o-Xylene	1.6
m-Xylene	1.65
1,4-Diethylbenzene	2.7
Propylene	3.4
Mesitylene	3.6
2-Methyl-2-butene	13.0

divided by the maximum oxidant for the appropriate toluene standard) or in toluene equivalents. Ten toluene standards were required to accommodate the chamber's variations over the past four years. At least three (and usually more) experiments were conducted with each individual compound, and the values given in Tables I to III represent the averages of each relative parameter. Even though chamber performance has varied, our results indicate that these relative values have remained fairly constant (with relative standard deviations generally within ±10%).

Table I presents the relative values for the maximum rate of nitrogen dioxide formation, Table II gives the maximum rates for hydrocarbon disappearance, and Table III presents the relative oxidant maxima. For comparison, we have included the values for α,α,α-trifluorotoluene, a particularly unreactive aromatic hydrocarbon; propylene, one of the principal hydrocarbons in automobile exhaust (13); and 2-methyl-2-butene, an especially volatile and reactive olefin that is often present in small amounts in gasoline (12).

Although these tables present only chemical measurements that need not be related directly to the unpleasant symptoms of photochemical smog (a fast rate of hydrocarbon disappearance is damaging only if the reaction products are harmful), a few generalizations about reactivity differences are in order. The reactivity differences between the compounds are large—a factor of 50 or more. Some compounds, like benzene and acetone, are essentially inert in the chemical parameters observed in this investigation, while other hydrocarbons (2-methy-2-butene) are highly reactive in all the chemical transformations observed in the irradiation chamber. Our results are in keeping with the usual hydrocarbon reactivity scales [see Dishart and Harris (14)] which suggest the following order of decreasing solvent reactivity (5, 14): (1) olefins, especially highly-substituted ones, (2) highly-substituted (trialkyl and dialkyl) benzenes, (3) monoalkylbenzenes, (4) alkanes (C_6 and greater), and (5) benzene. Carbon monoxide does not greatly affect relative reactivity rankings. For the chemical measurements and solvents in Tables I to III, the differences between oxygenated materials are less than those between hydrocarbons, and most of the oxygenated materials are much less reactive than the highly substituted olefins and aromatics. The individual reactivities of the various types of solvent materials are discussed in detail below.

Correlations between Various Measures of Photochemical Smog: The Lack of a Comprehensive Reactivity Scale. Photochemical smog formation can be evaluated by rate of formation, maximum values and dosages for smog products (i.e. nitrogen dioxide, ozone, aldehydes, and peroxyacyl nitrates), and by actual smog effects like eye irritation and plant damage. However, assessments of relative smog-forming reactivities are often based on only one measurement, usually the rate of NO_2 formation (1) or the extent of eye irriation (10) of photochemical smog. Correlations between chemical or kinetic measurements and actual smog symp-

Table III. Relative Oxidant Maxima in the Steel Chamber

Compound	Oxidant Maximum (relative to toluene)
Acetone	0.07
Benzene	0.2
Dimethylformamide	0.2
tert-Butyl alcohol	0.3
Trifluorotoluene	0.45
tert-Butyl acetate	0.5
Cyclopentanone	0.55
tert-Butylbenzene	0.55
tert-Amylbenzene	0.55
Cyclohexanone	0.6
Isopropyl alcohol	0.65
Indane	0.75
Ethyl acetate	0.8
n-Butylcyclohexane	0.8
Isopropyl acetate	0.85
n-Butyl acetate	0.85
n-Octane	0.85
Methyl ethyl ketone	0.9
p-Methane	0.95
n-Pentane	0.95
N,N-Dimethylacetamide	0.95
Ethyl alcohol	1.0
Toluene	1.00
Isobutyl acetate	1.0
Ethylbenzene	1.0
Isooctane	1.05
Diethyl ketone	1.05
n-Butylbenzene	1.2
sec-Butyl alcohol	1.25
Isobutyl alcohol	1.25
Methyl isobutyl ketone	1.35
1,4-Diethylbenzene	1.35
n-Butyl alcohol	1.4
Methyl n-butyl ketone	1.45
Tetrahydrofuran	1.45
4-Methyl-2-pentanol	1.45
o-Xylene	1.5
m-Xylene	1.7
Propylene	1.85
Diethyl ether	2.5
2-Methyl-2-butene	3.1
Mesitylene	3.2

toms are important. Other authors (4, 10) have evaluated linear correlation coefficients between a number of smog chamber parameters and observed several poor correlations (for example, eye irritation does not give a high linear correlation with any chemical measurements).

Such correlation coefficients measure only the linear relationship of two variables, and zero correlation does not necessarily imply independence (*15*). As checks of the actual relationships between the various chemical measures related to smog formation in our chambers, we plotted several pairs of variables. For those variables exhibiting some degree of linear relationship we calculated the tetrachoric correlation coefficients (*16*). A few of these correlation coefficients for the steel chamber are presented in Table IV, and Figure 1 shows a sample correlation plot comparing the maximum rates for hydrocarbon and nitrogen dioxide disappearance. Even though the correlation coefficient (0.74) for the variants in Figure 1 is relatively high, the scatter of the points indicates that a linear relationship of the two measurements is of limited value. Although we have reported a few examples of clear, nonlinear correlations between chamber parameters (*12*), our results generally support the observations (*4, 10*) of unimpressive correlations between various measures of smog formation. Clearly no single manifestation of photochemical smog provides an adequate appraisal of relative reactivity. As Tables I to III suggest, different chemical measurements result in somewhat different rankings for the relative reactivities of solvents, and rankings based on eye irritation, plant damage, or aerosol production also show different orders (*2, 4, 5, 10, 14*). A comprehensive scale for reactivities of solvent ingredients in smog formation would have to be based on a total reactivity index found by assigning appropriate weighting factors to the various smog manifestations and summing the weighted effects. Unfortunately, there is no sound, objective way to assign proper weighting factors to the various measurements, and no comprehensive solvent reactivity scale exists. With the variations encountered among irradiation chamber results (*12*) and the difficulties encountered in translating chamber observations to actual atmospheres, it is doubtful that an unimpeachable reactivity scale will soon be available.

However, studies of solvents with nitrogen oxides in irradiation chambers reveal mechanisms by which innocuous traces of materials can be transformed into photochemical smog. Useful generalizations have been derived from such investigations and are reviewed below.

Chemical Reactions and Intermediates Leading to Photochemical Smog. Since several excellent reviews of smog chemistry are available (*17, 18, 19*), an extensive discussion here of the chemical mechanisms involved in smog production is not necessary. The substances and intermediates available in smog for reactions with solvents include triplet atomic oxygen (from the photolysis of nitrogen dioxide); ozone; a number of peroxy-species (for example, peroxyacyl nitrates); and several types of radicals, such as ·OH, ·OOH, ·OR, ·OOR (where R is an alkyl or acyl group), NO, and NO_2. Many of these reactive substances are

Table IV. Sample Correlation Coefficients for Experiments in the Steel Chamber

Maximum Rate of NO_2 Formation and Other Variable	Correlation Coefficient
Average rate NO_2 formation	0.99
Rate NO_2 disappearance	0.62
Maximum rate hydrocarbon disappearance	0.74
Average rate hydrocarbon disappearance	0.78
Maximum formaldehyde	0.57
Time to maximum oxidant	−0.62
Maximum peroxyacetyl nitrate	0.4

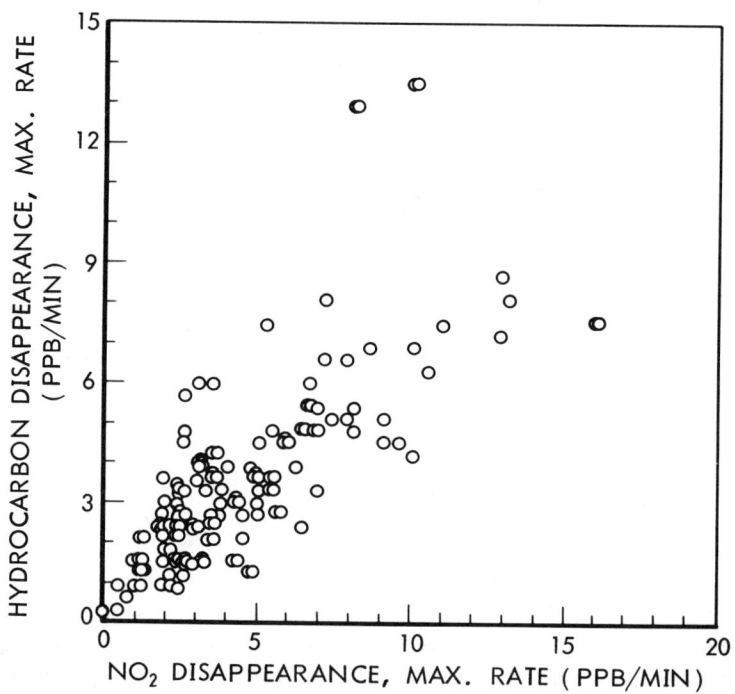

Figure 1. Sample correlation plot

strong oxidants and electrophiles, and the reactions of solvents in photochemical smog formation therefore often involve oxidation, displacement, or radical abstraction. Although the detailed mechanisms of the atmospheric reactions of solvents are not known, the following generalizations express the chemical measurements and results observed in several investigations.

Olefins. Olefins are the most reactive class of hydrocarbons in photochemical smog and have been studied extensively (1, 17, 18, 19). In general, as was perhaps first noted by Schuck and Doyle (20), the mechanism for olefin decomposition apparently involves electrophilic attack (by atomic oxygen, ozone, and other species) on the double bond. Thus, for most of the chemical reactions related to smog formation, olefin reactivity generally increases with additional alkyl (or other electron-donating) groups attached to the two carbon atoms joined by the double bond.

Aromatics. Although the complex processes by which aromatic compounds react in photochemical smog are not completely understood, the smog-forming tendencies of most aromatics can be rationalized by two initial modes of attack. For aromatic hydrocarbons having two or more alkyl groups on a benzene ring, a mechanism incorporating electrophilic attack on the aromatic ring as a rate-determining step explains the results from many smog chamber experiments (1, 4, 21, 22). For example, the enhanced reactivity of m-xylene, relative to its o and p isomers, is explained nicely by the ability of both methyl groups to donate electron density to the activated complex leading to some intermediate such as Structure I. With o- or p-xylene, only one methyl group could stabilize such intermediates.

$$\text{(I)}$$

Moreover, studies of products formed during the smog-chamber oxidations of multialkyl benzenes invariably indicate the presence of ring decomposition reactions (leading, for example, to the production of peroxyacetyl nitrate from m-xylene or mesitylene). Chemical reactivity of aromatic hydrocarbons is enhanced by an increasing number of alkyl groups, especially those in meta positions on the benzene ring.

For monoalkyl benzenes, there is evidence of a second mechanism for hydrocarbon decomposition. Since substantial amounts of benzaldehyde (representing a 5–25% yield) and traces (< 1% yield) of peroxybenzoyl nitrate have been identified (10) in smog chamber experiments with aromatic hydrocarbons having at least two benzylic hydrogens, hydrocarbon oxidation apparently may occur at the α-carbon atom. Application of the usual Hammett treatment and electrophilic substituent constants (23) to any of our rate measurements for toluene, *tert*-butylbenzene, and benzene indicates no linear Hammett relationship; toluene apparently reacts faster (by some other mechanism) than electrophilic substitution alone would predict.

However, experiments have shown that toluene-d_8 (completely substituted with deuterium) gives exactly the same smog chamber behavior as toluene itself under our standard conditions and measurements. Since a carbon–deuterium bond is stronger than a carbon–hydrogen bond, this lack of a deuterium isotope effect suggests that a carbon–hydrogen bond is not broken in the rate-determining steps of toluene's reactions.

Moreover, appreciable amounts (0.07 ppm maximum, 13% molar yield based on all the toluene) of peroxyacetyl nitrate (PAN) are formed in irradiation experiments with toluene. Heuss and Glasson (*10*) found 0.1 ppm (and also a 13% minimum yield) of PAN in their toluene experiments, and Altshuller *et al.* (*24*) obtained small amounts of PAN from several toluene–nitrogen oxide mixtures. In all our chambers, peroxypropionyl nitrate is observed as a product from ethylbenzene.

$$\text{C}_6\text{H}_5\text{-CH}_3 \xrightarrow[h\nu]{\text{NO}_x, \text{ air}} \text{CH}_3\text{C}(\text{O})\text{OONO}_2$$

$$\text{C}_6\text{H}_5\text{-CH}_2\text{CH}_3 \xrightarrow[h\nu]{\text{NO}_x, \text{ air}} \text{CH}_3\text{CH}_2\text{C}(\text{O})\text{OONO}_2$$

Clearly, monoalkylbenzenes may react by either side-chain oxidation or electrophilic attack on the benzene ring.

The photochemical reactivities of some substituted benzenes, like styrene whose high reactivity is derived from oxidation of the olefinic double bond, may reflect attack on the substituent itself rather than on the aromatic ring. However, to the extent that the susceptibility of the ring to electrophilic attack determines the reactivity of an aromatic, substituents would be expected to increase solvent reactivities in the following order: methoxy, ethoxy > alkyl (methyl > ethyl > isopropyl > *tert*-butyl) > unsubstituted > chloro > carbomethoxy, carboethoxy > cyano > nitro. This predicted order has not been thoroughly tested in irradiation chamber investigations. However, the Battelle study (*4*) found methyl benzoate to be quite unreactive, and our results show that α,α,α-trifluorotoluene (which can undergo neither electrophilic ring attack nor α-carbon abstraction reactions) is consumed even more slowly than benzene (*see* Table II).

Alkanes. In most of the chemical reactions observed in irradiation chambers, saturated hydrocarbons—even highly-branched ones such as p-menthane (1-isopropyl-4-methylcyclohexane)—have been quite unreactive. Since attack of alkanes by hydroxyl radical (26), atomic oxygen (27, 28), or ozone (29) follows the C–H reactivity order, tertiary > secondary > primary, the chemical measurements with alkanes would be expected to follow a clear pattern. However some alkanes (e.g., p-menthane) with tertiary hydrogens do not react more rapidly than those (e.g., n-octane) with only secondary and primary hydrogens, and hydrogen abstraction reactions often do not appear to be rate-determining steps.

Under reactive experimental conditions in smog chamber investigations (12, 25), alkanes give substantial amounts of oxidant. However, while Table III reveals that saturated hydrocarbons give about as much oxidant as many aromatics, there again is no evident correlation between oxidant production and the chemical structure of alkanes.

Alcohols. Like reactions of alcohols with atomic oxygen (30) or ozone (31), the mechanisms for decomposition of alcohols in smog chambers involve initial attack at a bond of the α-carbon atom. If there are no hydrogens on the carbon bearing the hydroxyl group (as in tert-butyl alcohol), the chemical reactivity is greatly reduced. Resonance interactions from the non-bonded p electrons of the oxygen atom favor hydrogen abstraction (or an insertion reaction) at the carbon atom to which the hydroxyl group is attached (32), and each alcohol (especially C_1–C_4) therefore generally reacts somewhat faster than the corresponding alkane.

Ethers. Solution chemistry suggests that ethers can oxidize by complex mechanisms involving initial attack at a carbon atom adjacent to the oxygen atom (33). Only two ethers (both with secondary hydrogens α to the ether group) were investigated in this study; both give fairly fast reactions. However, as with alcohols, blocking the α-carbon atom would presumably lower reactivity especially since tert-butyl ether does not react with ozone (34).

Acetates. All the acetates studied reacted slowly in our irradiation chambers. Just as tert-butyl alcohol is the least reactive alcohol, tert-butyl acetate is the least reactive ester considered here. Reactivity varies with the structure of the alcohol portion of the ester, and—as with alcohols and ethers—attack (hydrogen abstraction or displacement reactions) apparently occurs at the α-carbon of the alkoxy group. The ester acetyl group apparently is unreactive, but esters of higher acids might also undergo hydrogen abstraction reactions (favored by resonance stabilization by π-electron density of the carbonyl group) at the α-carbon of the acyl group.

Table V. Ketone Irradiations in the Small Glass Chamber

Ketone	NO_2 Average Formation Rate (ppb/min)	NO_2 Time for Maximum (min)	Ketone Consumed in 300 min, %	Maximum Values, Reaction Products [a] (ppb) PAN	PPN	tert-Butyl Nitrate
Acetone	1.3	>200	9–10	3–5	<1	<1
Methyl ethyl ketone	2.4	150	33	35	<1	<1
Diethyl ketone	4.2	112	27	36	18	<1
Methyl tert-butyl ketone	9.1	85	21	80	<1	44
Ethyl tert-butyl ketone	~8.5	80	22	23	27	40
Di-tert-butyl ketone	15.2	38	32	22	<1	100

[a] Analyzed by GLC with an electron capture detector; PAN = peroxyacetyl nitrate; PPN = peroxypropionyl nitrate.

Amides. The substantial increase in chemical reactivity on going from dimethylformamide to N,N-dimethylacetamide clearly implies reaction of the acetyl–methyl group. The virtual inertness of dimethylformamide indicates that electrophilic attack on the amide function is ineffective and that the N-methyl groups are not likely to react.

Ketones. The individual reactivities of ketones were measured in several irradiation chambers, including those of Battelle (4, 35) and LA-APCD (2, 3). In addition to the data given in Tables I to III for ketones in the steel chamber, we present Table V with experimental findings for ketone irradiations in the small glass chamber. Fairly good agreement exists between the relative chemical reactivities observed for several ketones in the different investigations.

However, the reactivity differences between ketones are oftentimes difficult to correlate with chemical structure. Acetone is most unreactive, implying that direct photolytic decomposition (17)

$$\mathrm{R\overset{O}{\overset{\|}{C}}R} \xrightarrow[(\sim 310 \text{ nm})]{h\nu} \mathrm{R\overset{O}{\overset{\|}{C}}{}^{\cdot} + R^{\cdot}}$$

is a minor process in smog chambers. With some ketones, γ-hydrogens may be abstracted internally by a photochemically excited carbonyl group (Norrish type II reaction), e.g.

$$\underset{RCH_2CH_2CH_2CR'}{\overset{O}{\|}} \xrightarrow{h\nu} RCH\underset{CH_2---CH_2}{\overset{H----O}{\diagup\diagdown}}C-R' \longrightarrow$$

$$RCH=CH_2 + CH_3\overset{O}{\underset{\|}{C}}R'$$

Any olefin formed by this process would enhance the conversion of NO to NO_2 and the build-up of ozone and aldehyde. This mechanism would explain, for example, Levy and Miller's observation (4, 35) that cyclohexanone (which cannot react in this way) reacts more slowly in a chamber than do other ketones with γ-hydrogens. However, irradiation in the steel chamber of methyl isobutyl ketone in air without nitrogen oxides results in little ketone disappearance (< 8% in 5 hours) or acetone formation, and direct photolysis of ketones is apparently not as important as the conventional (atomic oxygen, ozone, radical) smog reactions.

The surprisingly-rapid reactions observed for ketones containing *tert*-butyl groups (such as methyl *tert*-butyl ketone and di-*tert*-butyl ketone) have two implications:

(1) ketones do not react through the enol form, and (2) attack (such as hydrogen abstraction or displacement reactions) at the α-carbon, which appears to be important for many types of oxygenated compounds, does not explain the reactivity differences among ketones.

The chemical reactivities of ketones follow the order *tert*-butyl > ethyl > methyl, which is the stability order for the cations or radicals of these groups. This suggests that the mechanisms of ketone decomposition involve a rate-determining migration of an electrophilic alkyl moiety of the ketone, perhaps leading to the production of acyl and alkoxy radicals from some partially oxidized intermediate.

The production of peroxypropionyl nitrate from diethyl ketone and ethyl *tert*-butyl ketone *via* the conventional reactions

$$CH_3CH_2\overset{O}{\underset{\|}{C}}\cdot + O_2 \longrightarrow CH_3CH_2\overset{O}{\underset{\|}{C}}OO\cdot$$

$$CH_3CH_2\overset{O}{\underset{\|}{C}}OO\cdot + NO_2 \longrightarrow CH_3CH_2\overset{O}{\underset{\|}{C}}\diagdown OONO_2$$

is evidence for the formation of such acyl radicals from ketones. The tert-butyl nitrate obtained from tert-butyl ketones indicates the presence of the predicted tert-butoxy radicals and the reaction

$$(CH_3)_3CO\cdot + NO_2 \longrightarrow (CH_3)_3CONO_2$$

The mechanisms by which ketones react in smog chambers are particularly complex. Considerably more research, involving detailed product analyses and kinetic studies, is needed for a better understanding of ketone reactions in polluted air.

Conclusions

No single chemical or kinetic measurement accurately portrays the reactivity of substances in photochemical smog, but observations of the chemical reactions of solvent materials in irradiation chambers indicate how solvent components can react. Present knowledge of the atmospheric reactions of solvent ingredients is scanty and somewhat primitive, largely because of the formidable analytical problems involved in detecting and identifying low (ppb) concentrations of air contaminants. Several useful generalizations can be derived from the available data. Olefins and aromatics apparently are decomposed by electrophilic attack at unsaturated sites, and chemical reactivity therefore is raised by alkyl (or other electron-donating) groups on the double bond or ring. Alcohols, ethers, and esters apparently react at the α-carbon atom, and reactivity is reduced if no hydrogens are present on the α-carbon. The reactions of ketones are less understood than those of other solvents, but a rate-determining step may involve migration—as a radical or cation—of one of the alkyl groups attached to the carbonyl. The chemical reactivity of ketones may be increased by the presence of alkyl or other stabilizing groups on the carbon atoms adjacent to the carbonyl, and γ-hydrogens may also be involved in ketone reactions.

These generalizations should be tested and expanded in future research. It should be possible to evaluate, on the basis of chemical structure, the contributions that a solvent component may make to many of the chemical reactions that occur in urban atmospheres.

Acknowledgment

M. H. Grasley designed the steel chamber and performed the initial experiments. F. F. Farley, H. F. Richards, and J. E. Mahler have given advice and support to this project, and M. J. Hillyer provided computer programming and techniques for data analysis. R. A. Hurst, R. W. Franklin, J. T. Ronan, and L. R. Caputo have contributed valuable experimental assistance.

Literature Cited

1. Glasson, W. A., Tuesday, C. S., *Environ Sci. Technol.* (1970) **4**, 916.
2. Brunelle, M. F., Dickinson, J. E., Hamming, W. J., "Effectiveness of Organic Solvents in Photochemical Smog Formation," Air Pollution Control District, Los Angeles County, 1966.
3. Hamming, W. J., "Photochemical Reactivity of Solvents," SAE Paper 670809, Los Angeles, Oct., 1967.
4. Levy, A., Miller, S. E., "Final Technical Report on the Role of Solvents in Photochemical-Smog Formation," National Paint, Varnish and Lacquer Assoc., Washington, D. C., 1970.
5. Altshuller, A. P., *J. Air Pollut. Contr. Ass.* (1966) **16**, 257.
6. Stedman, D. H., Morris, E. D., Daby, E. E., Niki, H., Weinstock, B., "Abstracts of Papers," 160th National Meeting, ACS, Sept. 1970, WATR 26.
7. Wilson, W. E., Ward, G. F., "Abstracts of Papers," 160th National Meeting, ACS, Sept. 1970, WATR 28.
8. Bufalini, J. J., Gay, B. W., Kopczynski, S. L., *Environ. Sci. Technol.* (1971) **5**, 333.
9. Scofield, F., Levy, A., Miller, S. E., "Design and Validation of a Smog Chamber," National Paint, Varnish and Lacquer Assoc., Washington, D. C., 1969.
10. Heuss, J. M., Glasson, W. A., *Environ. Sci. Technol.* (1968) **2**, 1109.
11. Grasley, M. H., Appel, B. R., Burstain, I. G., Laity, J. L., Richards, H. F., "Abstracts of Papers," 158th National Meeting, ACS, Sept. 1969, ORPL 35.
12. Laity, J. L., Maynard, J. B., *J. Air Pollut. Contr. Ass.* (1972) **22**, 100.
13. Hurn, R. W., in "Air Pollution," A. C. Stern, Ed., Vol. III, Academic, New York, 1968.
14. Dishart, K. T., Harris, W. C., *Proc. Div. Refining, Amer. Petrol. Inst.* (1968) **48**, 636.
15. Hahn, G. J., Shapiro, S. S., "Statistical Models in Engineering," pp. 63–65, John Wiley and Sons, New York, 1967.
16. Dixon, W. J., Massey, F. J., Jr., "Introduction to Statistical Analysis," 2nd ed., pp. 267–269, McGraw Hill, New York, 1957.
17. Leighton, P. A., "Photochemistry of Air Pollution," Academic, New York, 1961.
18. Altshuller, A. P., Bufalini, J. J., *Photochem. Photobiol.* (1965) **4**, 97.
19. Altshuller, A. P., Bufalini, J. J., *Environ, Sci. Technol.* (1971) **5**, 39.
20. Schuck, E. A., Doyle, G. J., Report No. 29, Air Pollution Foundation, San Marino, Calif., 1959.
21. Kopczynski, S. L., *Air Water Pollut.* (1964) **8**, 107.
22. Levy, A., Miller, S. E., Ann. Meet. Air Pollut. Contr. Ass., 62nd, New York, 1969, Paper 69-13.
23. Brown, H. C., Okamoto, Y., *J. Amer. Chem. Soc.* (1958) **80**, 4979.
24. Altshuller, A. P., Kopczynski, S. L., Lonneman, W. A., Sutterfield, F. D., Wilson, D. L., *Environ. Sci. Technol.* (1970) **4**, 44.
25. Altshuller, A. P., Kopczynski, S. L., Wilson, D., Lonneman, W., Sutterfield, F. D., *J. Air Pollut. Contr. Ass.* (1969) **19**, 787.
26. Greiner, N. R., *J. Chem. Phys.* (1967) **46**, 3389.
27. Herron, J. T., Huie, R. E., *J. Phys. Chem.* (1969) **73**, 3327.
28. Wright, F. J., *J. Chem. Phys.* (1963) **38**, 950.
29. Hamilton, G. A., Ribner, B. S., Hellman, T. M., Advan. Chem. Ser. (1968) **77**, 15–25.
30. Avramenko, L. I., Kolesnikova, R. V., in *Advan. Photochem.*, W. A. Noyes, Jr., G. S. Hammond, J. N. Pitts, Jr., Eds., (1964) **2**, 25–62.

31. Whiting, M. C., Bolt, A. J. N., Parish, J. H., Advan. Chem. Ser. (1968) **77**, 4–14.
32. Tedder, J. M., *Quart. Rev.* (1960) **14**, 336.
33. Bailey, P. S., *Chem. Rev.* (1958) **58**, 925.
34. Price, C. C., Tumolo, A. L., *J. Amer. Chem. Soc.* (1964) **86**, 4691.
35. Levy, A., Miller, S. E., "Abstracts of Papers," 158th National Meeting, ACS, Sept. 1969, ORPL 36.
36. Calvert, J. G., Pitts, J. N., Jr., "Photochemistry," p. 382, John Wiley and Sons, New York, 1967.

Received March 14, 1972.

8

A Practical Approach to Solvent Applications in Coatings and Inks

D. K. SAUSAMAN

Commercial Solvents Corp., 1331 S. First St., Terre Haute, Ind. 47808

> *The selection of solvents and solvent blends for use in coatings and inks is based upon solubility/viscosity characteristics and application/performance properties. Published solubility parameters and hydrogen bonding indexes are used to construct two-dimensional solubility maps. Methodology is described, and illustrations are shown. Data are provided on evaporation times of neat solvents, viscosities and dry times of polymer solutions, electrostatic characteristics of solvents, and on selected solvent blend recommendations for several polymers. Unpublished test methods for flow testing and for substrate testing are provided. Combination of the results from these areas provides a viable method for practical solvent blend selection; this approach is faster than random trial-and-error and can result in superior, formulated solvent blends.*

Almost every area of the coatings and inks industry has changed technologically in the past ten years. Applications technology has a new sophistication that can be seen in the copious data on solubility parameters, evaporation rates of mixed solvents (1), and evaporation rate measurements using radio-tagged solvents (2) and other techniques. However, new regulations on solvent emissions have met the new technology. This paper presents several approaches to various aspects of solvent formulation, and it develops data which can be used to achieve functional properties in coatings and inks.

Resin examples are limited to those polymers which either require or are benefited significantly by oxygenated solvents. Data are categorized into solubility/viscosity and application/performance properties. Specific test methods are included as needed.

Discussion

Solubility/Viscosity. The term "solvency" requires a simplified definition. Solvency is the ability of a compound (solvent) to allow molecular separation and miscibility of another compound (solute) within it. When this occurs, a solution is formed. Where solution does not occur, there may be too much or too little solute present.

Solubility parameters and the accompanying hydrogen bonding indexes express certain measurable properties of solvents, but how do these values influence day-to-day practical formulation? The answer can be found by constructing a two-dimensional solubility map and placing on it the solubility parameters and hydrogen bonding indexes for a group of solvents (3). (There are varying values reported by several sources. All the values used should be from the same source to prevent false assumptions by mixed data input, and when available, those recommended by the resin supplier are good for a start.) The specific map locations of the solvents are based on published solubility parameter numbers (δ) and hydrogen bonding indexes (γ).

For unknown resins, use several solvents and/or blends which appear at varying points on the chart (*see* Figure 1). Determine by observation which solvents will dissolve the resin. The non-volatile content should be kept low at the outset so that solubility is not masked by an overload of solute. By increasing the level of solute, the capacity of the solvent for the resin is observable. The viscosity at any solute concentration can be measured, and a cost calculation of the solvents can be made. Using

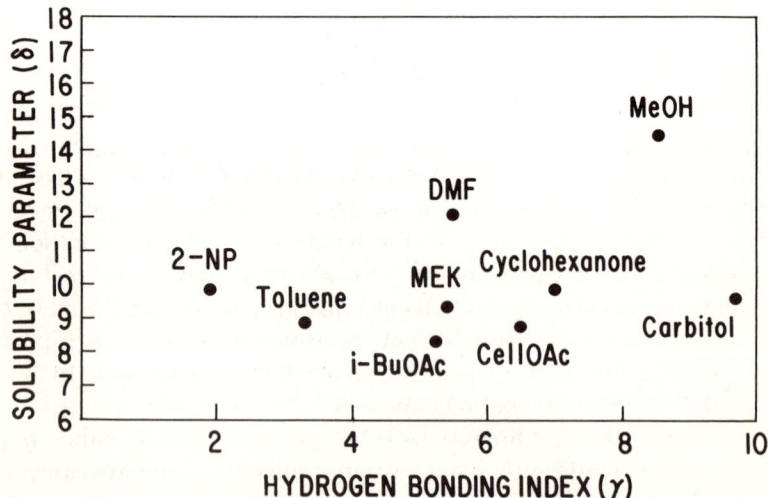

Figure 1. Solvent point plots

Table I. 20% Vinyl Resin

Soluble	δ	γ	Soluble	δ	γ
Acetone	10.0	5.9	Ethyl benzoate	8.2	5.5
Benzonitrile	8.4	2.5	Isobutyl acetate	8.3	5.5
n-Butyl acetate	8.5	5.5	Isophorone	9.1	7.0
Butyronitrile	10.5	2.5	Isopropyl acetate	8.4	6.0
Chloroform	9.3	2.5	Mesityl oxide	9.0	5.5
Cyclohexanone	9.9	7.0	Methylene chloride	9.7	2.5
Cyclopentanone	10.4	5.5	Methyl benzoate	10.5	5.5
Diacetone alcohol	9.2	6.8	Methyl ethyl ketone	9.3	9.3
Diethyl ketone	8.8	6.2	Methyl isoamyl ketone	8.4	5.5
Diisobutyl ketone	7.8	4.8	Methyl isobutyl ketone	8.4	6.2
Dimethyl formamide	12.1	5.5	Nitrobenzene	10.0	2.9
Dimethyl sulfoxide	12.0	5.5	1-Nitropropane	10.3	1.9
1,4-Dioxane	9.9	6.7	2-Nitropropane	9.9	1.9
Ethylene dichloride	9.8	2.5	Piperidine	8.7	8.5
Ethyl acetate	9.1	5.2	Propylene oxide	9.2	5.5
			Pyrridine	10.7	8.5

Partially Soluble	δ	γ	Partially Soluble	δ	γ
Amyl formate	8.2	5.5	Dioctyl phthalate	7.9	5.5
Aniline	11.8	4.5	Ethyl glycol monobutyl ether	8.9	7.0
n-Butyl bromide	8.7	5.5			
Isobutyl butyrate	7.8	5.5	Ethyl glycol monoethyl ether	9.9	6.8
n-Butyl butyrate	8.1	5.5			
Carbon tetrachloride	8.6	3.4	Ethyl glycol monothyl ether acetate	8.7	6.5
Chlorobenzene	9.5	2.5			
Diethyl carbonate	8.8	5.5			
Diethyl glycerol monoethyl ether acetate	8.5	5.5	Ethyl lactate	10.0	5.5
			Methyl nonyl ketone	7.8	5.5
			Propylene carbonate	13.3	5.5

Insoluble	δ	γ	Insoluble	δ	γ
Acetonitrile	11.9	2.5	Ethylbenzene	8.8	2.5
Benzene	9.2	2.6	2-Ethyl hexanol	9.5	8.5
n-Butyl alcohol	11.4	8.5	n-Heptane	7.4	2.2
Carbon disulfide	10.0	2.5	n-Hexane	7.3	2.1
Cyclohexane	8.2	2.2	Isopropyl alcohol	11.5	8.7
Diethyl glycol monobutyl ether	8.9	6.8	Methanol	14.5	8.5
			Methyl isobutyl carbinol	10.0	8.5
Diethyl glycol monoethyl ether	9.6	9.7	Nitromethane	12.7	2.5
Dipentine	8.5	2.5	Toluene	8.9	3.3
Ethylene glycol	14.6	8.6	VM&P naphtha	7.6	2.5
Ethyl alcohol	12.7	8.6	Xylene	8.8	3.5

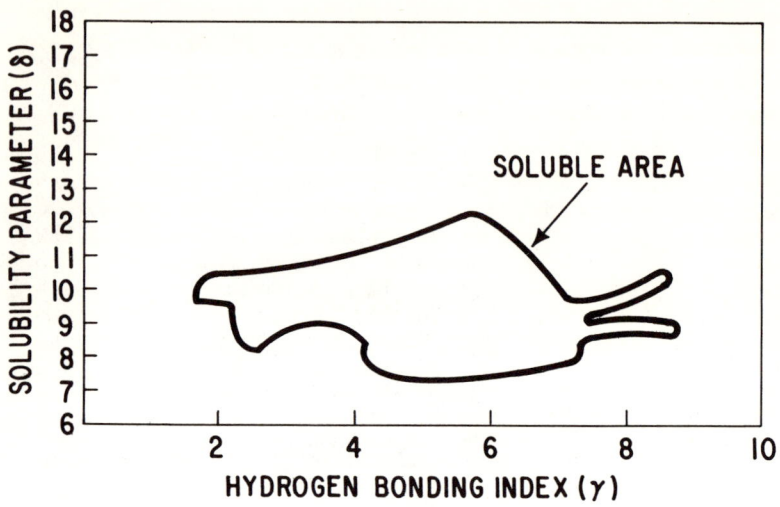

Figure 2.   VYHH solubility map

Table II.   Aliphatic/Alcohol Tolerance—30 Day Viscosities 20% VYHH Solutions[a]

VM&P Naphtha

	%Dilution			
	0	10	20	30
2-Nitropropane	305	305	576	Gel
Methylene chloride	480	950 (stringy)	Gel	Gel
MIBK	295	490	Gel	Gel

99% Isopropyl Alcohol

	%Dilution			
	0	10	20	30
2-Nitropropane	310	362	641	Gel
Methylene chloride	465	1070 (stringy)	Gel	Gel
MIBK	290	505	769	Gel

[a] *Formula:* Resin 20%, active solvent 40%, xylene 40%.

these data, the number of further trials required to define the desired system is reduced.

In many cases, limits exist which also narrow the search. There may be a maximum limit on solvent cost or viscosity, a minimum on nonvolatile content, and so on. As an example, a vinyl resin copolymer (Bakelite vinyl resin VYHH) was evaluated for solubility. The only requirement was fluidity at 20% non-volatile (wt) content. The solvents used to evaluate solubility are shown in Table I, and a two-dimensional solubility map is shown in Figure 2. (This is a local map for one resin at one non-volatile content.) Looking at this map, a question arises regarding blends of solvent and non-solvent which fall within the solubility limits. (Improved economics are normally the reason for using non-solvents such as hydrocarbons.) Again with VYHH as the example, blend solubility is plotted. Table II lists the viscosities of several solutions

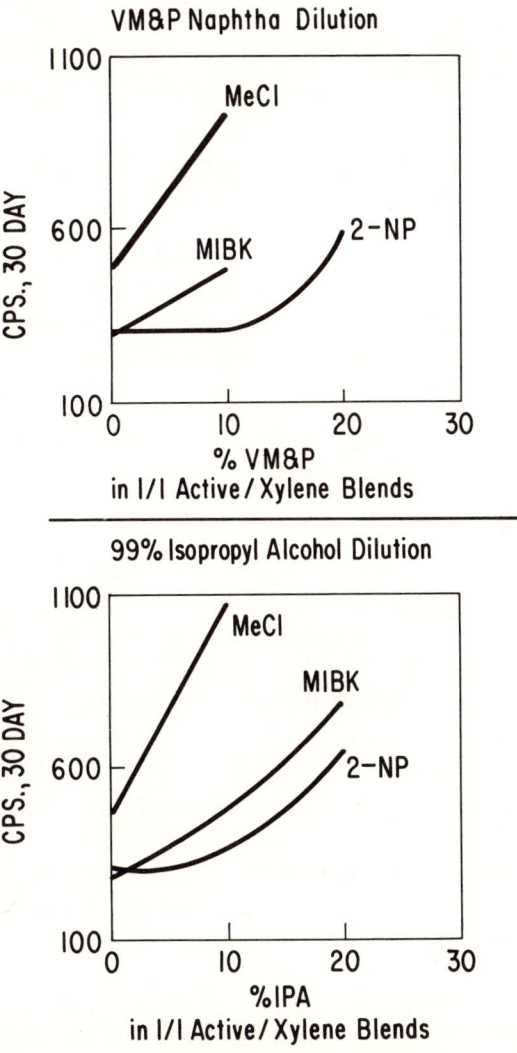

Figure 3. *Effect of non-solvent usage in VYHH solutions*

which are plotted in Figure 3. Figure 4 shows that a viscosity *vs.* economics plot can visually demonstrate the best compromise between these two factors. Normally, the non-solvents closest to the mapped solubility limits are better tolerated than those further away. The map locations of blends can be determined by measuring the distance between solvent points and plotting the point which represents the ratio of solvents in the blend.

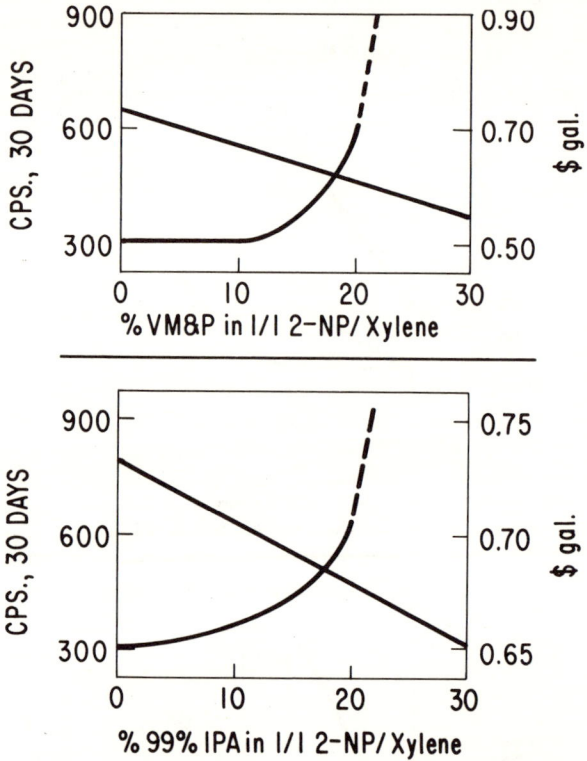

Figure 4. *Optimum viscosity/cost determination*

Viscosity measurements can be made using several different methods and attendant equipment. To help relate viscosities reported in different units, Table III reports comparative viscosities using different methods of measurement. The accuracy will lessen as non-Newtonian flow (thixotropy, pseudoplasticity, etc.) increases, but it serves as an excellent guide.

Application/Performance Properties. After the solubility/viscosity properties of a solution or a group of solutions have been determined, application and performance properties must be evaluated. In many cases, a close tie between these properties exists. Obviously, some thought has been given to limiting the potential solvents evaluated to get to this point. For example, these limits would preclude an extremely fast-evaporating solvent for a baking application or a water-soluble solvent where quick water resistance is required. Some of the properties of interest in coatings and inks are: (1) dry time, (2) flow, (3) solvent retention, (4) gloss, (5) effect on substrate, (6) build-up on application rollers, and (7) cobwebbing on spray application. These factors are tied together in several places; altering one characteristic may alter others.

Dry times may be evaluated by applying coatings of equal volume non-volatile content at equal film thicknesses and then timing to determine dry. This measurement can become somewhat subjective as operators change, but it is useful as a comparative method. Mechanical equipment is now available for this measurement. Also, more sophisticated information which involves evaporation rate analysis (ERA), escaping coefficients of solvents, and evaluation of solvent release by radioactive tagging of solvents has appeared within the past 18 months in the *Journal of Paint Technology, American Ink Maker,* etc. From a practical standpoint, when faster dry is required, solvents of higher evaporation rates are incorporated, and slower evaporating solvents are deleted. Table IV lists the evaporation times of many common solvents (4).

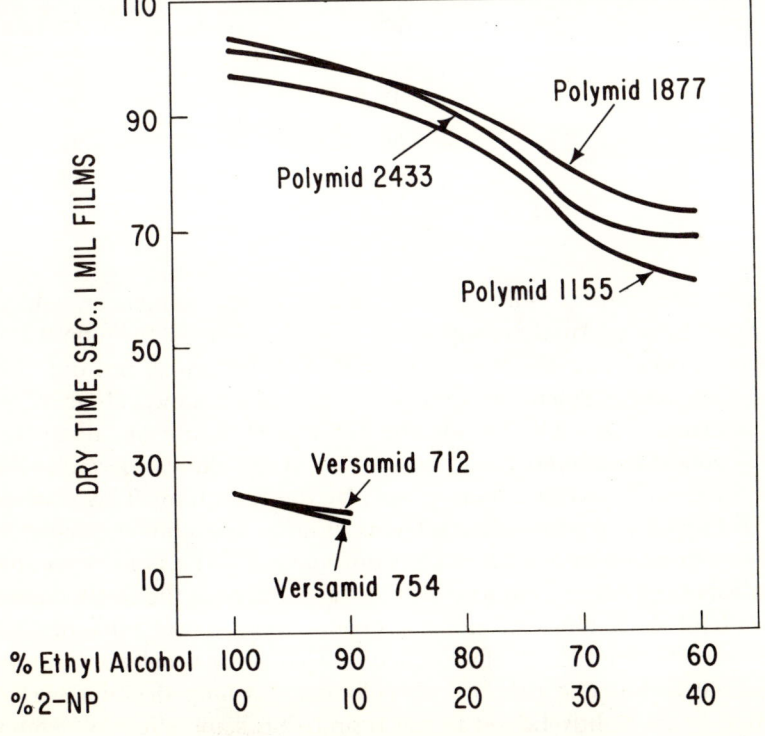

Figure 5. Dry times of polyamide inks

Table III. Viscosity

Zahn Viscosity Cup

#1	#2	#3	#4	#5
36.	—	—	—	—
43.5	—	—	—	—
55.5	23.	—	—	—
65.5	27.	—	—	—
78.5	32.5	—	—	—
—	37.5	—	—	—
—	46.	—	—	—
—	51.	—	—	—
—	60.	—	—	—
—	—	25.5	—	—
—	—	27.5	20.5	—
—	—	29.5	21.5	—
—	—	31.5	22.5	—
—	—	33.5	24.5	16.5
—	—	35.	25.5	16.5
—	—	37.	27.	17.5
—	—	39.	28.5	18.5
—	—	41.5	30.5	20.
—	—	44.5	32.	21.5
—	—	47.5	34.5	22.8
—	—	50.	36.	24.
—	—	54.	39.	26.
—	—	60.	43.5	29.5
—	—	—	50.5	34.5

Evaporation rate is not the only factor which influences solvent release (dry time). Another important consideration is the affinity of the solvent molecule for the resin molecule. This influences not only dry time but also solvent retention, film porosity, and to some extent, effect on the substrate. Low affinity for the resin is certainly an advantage added to economics offered by a non-solvent. Low affinity permits relatively complete release of the non-solvent by the resin film. This concept is of major importance with active solvents. Figure 5 exhibits the typically reduced dry times observed when 2-nitropropane (2-NP) is incorporated into an alcohol-soluble polyamide ink (5). Even though the slower evaporation time of the 2-NP (415 min) is greater than that of ethanol (328 min), the dry time is drastically reduced. This is an excellent example of the benefit of keeping the film open longer (to help drying) with a solvent that is not tightly bound to the resin. This same effect is demonstrated in thin film vinyl coatings by the data in Figure 6. Newer meas-

Conversion Chart

Ford #4 Cup, sec.	Gardner-Holdt Bubbles	Centipoises	Krebs Units
13.6	A-2	—	—
15.3	A-1	—	—
19.	A	40	—
22.	B	—	—
27.	C	—	—
30.	D	100	—
36.	E	—	—
40.	F	—	—
46.	G	—	—
55.	H	—	—
62.	I	200	—
68.	J	—	—
74.	K	—	59
81.	L	300	61
86.	M	—	62
91.	N	—	63
99.	O	—	64
107.	P	400	65
116.	Q	—	66
125.	R	—	67
133.	S	500	68
146.	T	550	69
167.	U	600	71
199.	V	900	78

urement of retained solvent using gas chromatographic techniques clearly demonstrates this same point. When a resin can be dissolved in a blend containing solvents which have low resin affinity, both total volatile release and release of the active solvent are improved. The newer gas chromatographic techniques reveal not only quantity, which formerly was determined by weight (gravimetric analysis) but also reveal which solvents remain (6). It is premature to draw definite conclusions, but preliminary work suggests that the order of addition of solvents can also strongly influence the type and quantity of solvent retained.

A surprising facet of dry time and flow has appeared in several systems (7). Data indicate that flow and dry time are not necessarily in direct proportion. The wetting effects of the solvents also can influence flow. Returning to the polyamide inks, a flow test (described in Appendix A) was conducted on many of the inks. Evaluation of the flow in a pair of systems exhibiting 97-sec dry (ethanol) and 61-sec dry (60/40 ethanol/

Table IV. Evaporation Times of Several Commercial Solvents (9)
Shell Thin Film Evaporometer, 25°C, 0% Relative Humidity

Solvent	Evaporation time, min.
Acetone	82
n-Butyl alcohol	1080
Cyclohexanone	1570
Diacetone alcohol	3840
Diethylene glycol monobutyl ether	50000
Diisobutyl ketone	2440
1,4-Dioxane	70
85–88% Ethyl acetate	115
190 Proof ethyl alcohol	328
Ethylene glycol monoethyl ether acetate	2530
Isobutyl acetate	305
Isobutyl alcohol	740
Isophorone	20000
Methyl alcohol	221
Methyl ethyl ketone	121
Methyl isobutyl ketone	282
2-Nitropropane	415
85–88% Isopropyl acetate	134
n-Propyl acetate	220
Toluene	230
Xylene	760

Table V. Polyamide Inks

Polymid 1877 [a]	Anhydrous Ethanol/2–Nitropropane				
	100/0	90/10	80/20	70/30	60/40
Viscosity, cps	23	21	19	24	27
1 mil film dry time, sec	101.5	98	91	79.5	72.5
Flow distance, mm	43.5	46	47	48	49

[a] Lawter Chemical Co. 25% non-volatile by weight solutions.

2-NP) demonstrates an inverse dry time/flow relationship (Table V). The system which dries fastest also exhibits the best flow. Flow (or leveling) is usually subjectively measured by gloss or by observation. A new technique uses the New York Paint Committee leveling bar. This simple technique allows a better comparison than the older, subjective methods.

When blending solvents for any application, the last volatile component to leave the film should be a solvent. If this precaution is neglected, the coating or ink can easily end up with a grainy or blushed, low gloss appearance from resin particle precipitation.

Another important consideration, particularly in inks, is that the solvent system should be strong enough to redissolve partially-dried ink from the application rolls. This prevents build-up on the rolls and prevents fuzzing of the fine print. There are no tests to define this property other than press runs.

Evaporation rates of solvents play a large role in air spray application. If the solvent is too fast, too much solvent leaves the coating prior to contact with the substrate. This causes stringing or cobwebbing of the spray or rough surfaces and orange-peeling on the work being sprayed. If the solvent is too slow, the coating will tend to sag or run.

Another technological area is electrostatic application of coatings or inks. Table VI lists a number of common solvents with their electrical resistivities, relative evaporation rates, approximate costs, and a summary comparing the most logical choices. Where adjustment is required for

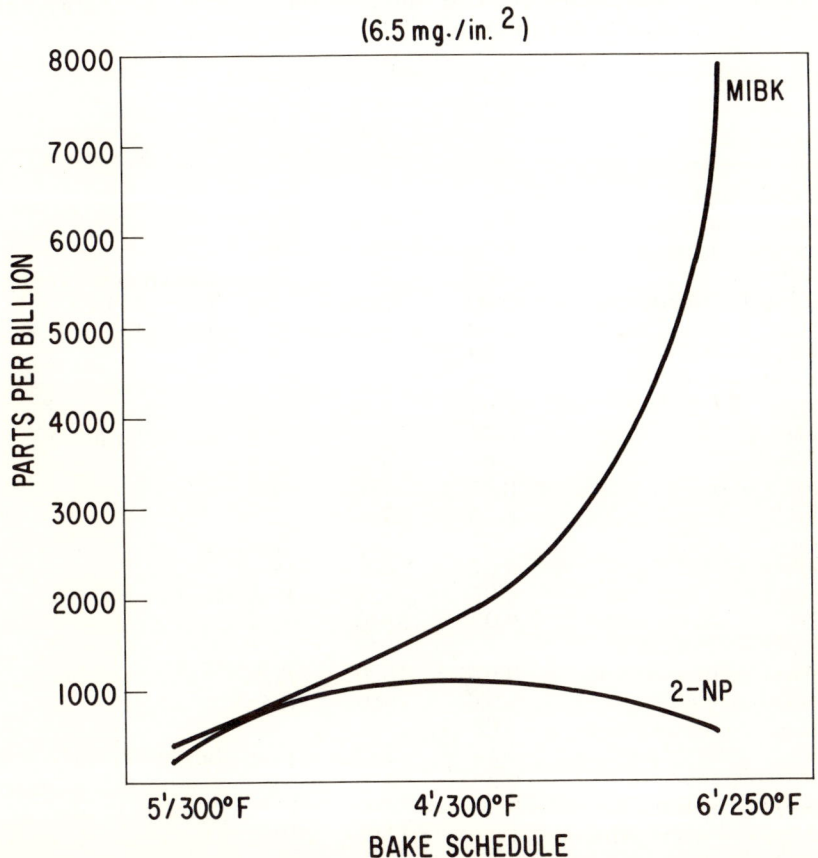

Figure 6. Solvent retention in vinyl films (6.5 mg./in.²)

proper application using electrostatic equipment, these data and Table VII will aid one in choosing a proper solvent or blend. A word of caution is needed here: do not test unpigmented coatings and expect duplicate conductivity when pigment is added (8).

In choosing a solvent blend, its effect on a soluble substrate is an important consideration. An example which illustrates this is the application of a vinyl copolymer ink to a low-molecular-weight polyvinyl chloride (PVC) film. The Commercial Solvents Corp. test for determining film substrate attack is given in Appendix B. This test method could be combined with an Instron to provide more exact detail on film strength loss. As shown by the data in Table VIII:

(1) Toluene was the solvent which weakened the film even though it is not itself a solvent for PVC;

Table VI. Suitability of Polar Solvents for Electrostatic Spraying Adjustment in Order of Increasing Resistivity

	Resistivity (megohms)	Evaporation Time, sec.[a]	Approx. Cost per lb, cents	Summary
Nitromethane	0.02	—	31	
Nitroethane	0.02	—	33	
Dimethyl formamide	0.02	2280	22	
Methanol	0.02	221	4.1	too volatile, etc.
Shellacol (anhydrous)	0.02	328	9.2	Best of alcohols
Methyl Cellosolve	0.02	885	15	Best of glycol ethers
Shellacol 190	0.03	328	8	
2-Nitropropane	0.03	415	17	
Fotocol AA–1	0.04	328	9.2	
Acetone	0.04	82	5.9	too volatile
Cellosolve	0.04	1210	16.5	
Butyl Cellosolve	0.05	6750	17	
Isobutyl alcohol	0.15	740	8	
Isopropyl (99%)	0.07	319	7	
Isophorone	0.07	20000	18	
Methyl ethyl ketone	0.07	121	10.5	best of ketones
Diacetone alcohol (acetone free)	0.075	3840	13.5	
Methyl isoamyl ketone	0.075	1020	17.5	
n-Butyl alcohol	0.08	1080	12	
Methyl n-butyl ketone	0.08	—	—	
Methyl isobutyl ketone	0.13	282	13.5	better choices
Cyclohexanone	0.20	1570	18	available in same
Diisobutyl ketone	0.42	2440	15	solvent class
Methyl isobutyl carbanol	1.4	1710	16	
Tetrahydrofuran	2.2	—	36	

[a] Shell thin film evaporometer.

(2) After film penetration was determined, the best compromise of dry time, economics, and viscosity was picked.

Even though the relative evaporation rate of the recommended blend is considerably faster than the standard, flow was greatly improved. The improved flow eliminated voids in the print and improved print quality.

Another consideration is air pollution. Rule 66 (Los Angeles Air Pollution Control District's emissions law) is normally the guideline because it is the most limiting, and the coming national air pollution standards will follow these lines closely. Formulating for compliance to

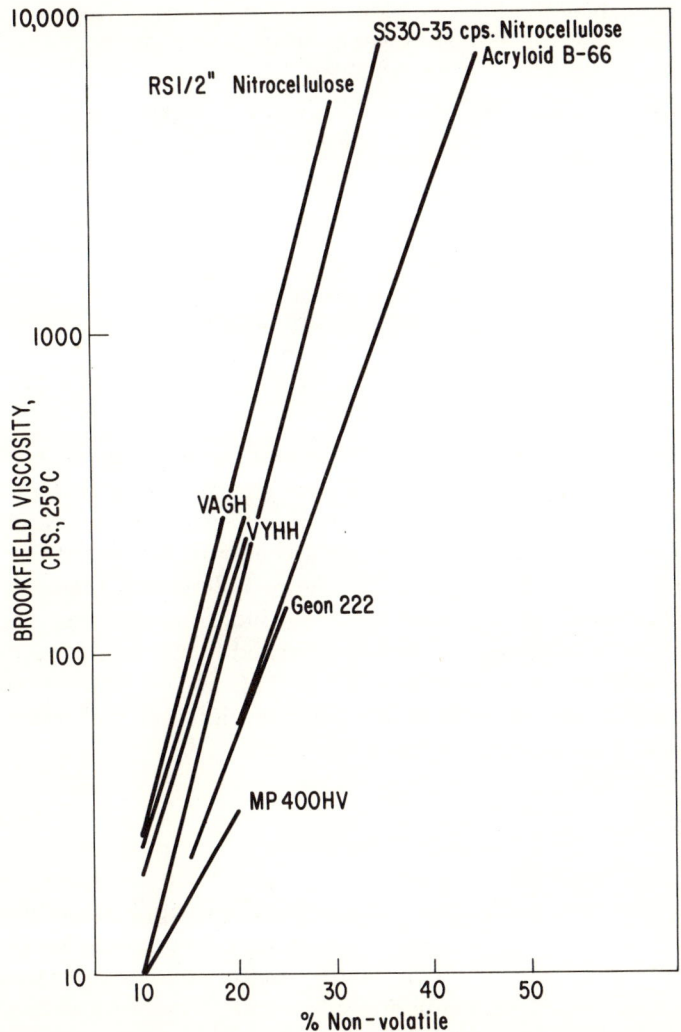

Figure 7. Solutions in universal solvent

Table VII. Best Polar Solvents for Electrostatic Spraying Adjustment

	Strengths	Weaknesses
Dimethyl formamide	lowest resistance in resin systems.	cost: incompatibility, especially min. spirits.
Methyl Cellosolve	lower resistance in some resin systems.	some incompatibility; more toxic than NP's.
Anhydrous Shellacol	low cost; lower resistance in many resin systems.	incompatible with many resins; high volatility.
Methyl ethyl ketone	effectiveness in lowering resistance on a par with the NP's; lower cost.	high volatility.
2-Nitropropane	good resistivity reduction; med. evap. range; good compatibility in a broad range of systems.	intermediate in cost.

air pollution regulations simply involves setting up limits on usable solvents and their allowable content. Proceed, considering these limits, as described earlier to optimize other requirements. Table IX defines a general purpose, exempt solvent blend and reports the viscosities of several polymer species dissolved in it. Figure 7 illustrates these viscosities graphically. Speed of dry, solvency, economics, application, and other characteristics of this blend are easily adjusted by manipulating the components to achieve the needed properties. This may be done by adjusting the ratios of the components or by selecting different molecular weight homologs of the components for use.

Summary

Using the guidelines set forth earlier and taking advantage of new data and techniques as they evolve enables proper application of solvents in coatings and inks. Older test methods may incorporate newer, more sophisticated measurement techniques for better definition. Experimental trials can be lessened, and superior solvent blends can be prepared. As data evolve for several potentially useful systems, a weighted ranking value can be assigned to make the proper selection.

Table VIII. Formulations and Data

	Recommended		−1		−2	
	Pounds	Gallons	Pounds	Gallons	Pounds	Gallons
2-Nitropropane	69.3	8.41	66.1	8.02	69.3	8.41
Fotocol AA–1	122.3	18.50	41.3	6.55	122.3	18.50
Acetone	142.6	21.65	138.3	20.99	71.8	9.38
Cyclohexane	122.3	18.75	123.9	19.00	122.3	18.75
Isopropyl acetate	236.4	32.69	165.2	22.84	321.4	44.96
Toluene	—	—	165.2	22.90	—	—
Total	692.9	100.00	700.0	100.00	697.1	100.00

	Isopropyl Acetate	Recommended	−1	−2
Weight./gallon, lbs	7.23	6.93	7.00	6.97
Dollars/gallon[d]	.8315	.6074	.5050	.6607
Dollars/lb[d]	.1150	.0877	.0722	.0943
Calculated relative evaporation rate	439	525	492	463
Tack-free time, min.	25	19–20	N/R[a]	23
Viscosity, cps., 77°F	108	57	N/R[a]	83
Film penetration time,[b] min.	[c]	[c]	20	[c]

[a] N/R = Not reported.
[b] Quadruplicate runs.
[c] Test stopped at three minutes. Bar could not be easily pushed through film after three minutes.
[d] Approximate.

It sometimes seems impossible to visualize assimilation of new techniques, but bit by bit they come into common usage. As scientists and/or practitioners of the art, it is incumbent on the formulator to maintain an open mind regarding new techniques. As technology becomes more sophisticated, the areas which are adaptable to the practical approach should be brought into use. Only by rigorous evaluation can the merits of such new approaches be determined.

Appendix A: Testing Flow

Flow/Wetting Properties. Flow/wetting improvement was evaluated by placing 0.1 ml of each of the systems on a glass plate supported at a 45° angle and allowing them to run down the glass. The length of the runs is a measure of the wetting of the glass and the flow of the ink.

Table IX. Universal Blend

	Pounds	Gallons
Acetone	200	30.0
Cyclohexane	130	20.0
Toluene	144	20.0
Isopropyl acetate	108	15.0
2-Nitropropane	82	10.0
99% Isopropyl alcohol	33	5.0
Total	697	100.0

RS 1/2" Nitrocellulose (30% Ethanol)						
% Non-volatile, wt.	10.1	15.7	20.2	25.3	29.9	35.0
1 day viscosity, cps	26	123	515	1989	5203	>10000
SS30-35 cps Nitrocellulose (30% IPA)						
% Non-volatile, wt.	10.1	15.7	20.2	25.3	29.9	35.0
1 day viscosity, cps	9	43	187	730	2604	7844
100% Acryloid B-66 Acrylic Resin						
% Non-volatile, wt.	20	25	30	35	40	45
1 day viscosity, cps	6	20	37	82	335	776
Bakelite VMCH Vinyl						
% Non-volatile, wt.	10	15	20			
1 day viscosity, cps	51	58	233			
Bakelite VYHH Vinyl						
% Non-volatile, wt.	10	15	20			
1 day viscosity, cps	20	60	230			
Bakelite VAGH Vinyl						
% Non-volatile, wt.	10	15	20			
1 day viscosity, cps	22	69	267			
Geon 222 Vinyl						
% Non-volatile, wt.		15	20	25		
1 day viscosity, cps		23	64	140		
Vinoflex MP400 HV Vinyl						
% Non-volatile, wt.	10	15	20			
1 day viscosity, cps	8	17	32			

Note that the viscosities and non-volatile contents of the systems should be equal. The viscosities of several pigmented systems studied are essentially equal, but the 2-NP systems dry faster than the 100% alcohol standard. If the systems wetted the substrate equally, the 2-NP systems would flow less because of their faster dry. However, the 2-NP systems flowed further than the standard system. Thus the 2-NP polyamide system wets the substrate better than the standard polyamide system and exhibits better flow even though the dry is faster.

Flow was also studied qualitatively in pairs by simultaneously drawing them down on polypropylene film with a #50 wire wound rod. The leveling was visually evaluated. The 2-NP systems were both superior to the 100% alcohol system.

Appendix B: Film Substrate Attack by Solvents, Inks, and Coatings

This is a qualitative test to evaluate differences in the effect of various solvents, solvent blends, inks, and coatings on films which are subject to attack (*i.e.*, PVC films).

Secure a piece of the test film over a 3½ inch orifice (a standard, wide-mouth 16 oz. jar with the center of the lid cut out to leave only the rim is a good, inexpensive piece of equipment). Allow a slight ($\sim$¼") depression in the center of the film so that the material to be tested does not spread excessively.

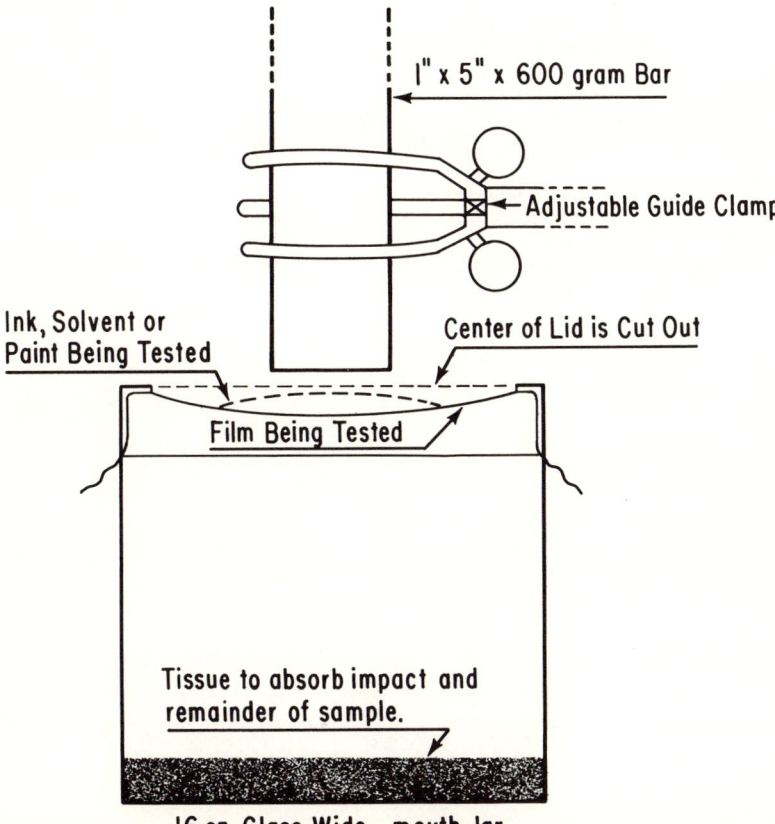

Figure B1. Rough sketch of film penetration test procedure as described

Place a steel bar (we use one which is 5" long, 1" in diameter, weighing approximately 600 grams) in an adjustable clamp immediately over the center of the film.

Put 2 cc of the test fluid on the film and immediately place the bar on the film. (Do not drop the bar onto the film.) The clamp must be such that the bar is guided vertically, but not impeded. Determine the time required for the bar to penetrate the film. Duplicate trials should be made. Blow test jar free of solvent between trials.

It has been found that rapidly-volatilizing systems will cool the bar and thus influence time-to-penetration. To avoid this, the bar should be at ambient temperature for each trial. Where heavy substrates or weak solvents are being tested, bar weight may be increased to facilitate quicker testing.

Literature Cited

1. *J. Paint Technol.* (1969) **41**, 692.
2. Borer, W. Z., *et al.*, *J. Paint Technol.* (1971) **43**, 61.
3. E. I. DuPont de Nemours & Co., Electrochemicals Department, "A New Dimension in Solvent Formulation," Wilmington, 1965.
4. Nelson, R. C., *et al.*, *J. Paint Technol.* (1970) **42**, 644.
5. Commercial Solvents Corp., "2-Nitropropane in Polyamide Inks," *Tech. Bull.* (1970) **29**.
6. Sausaman, D. K., unpublished data.
7. Bennett, R. R., *Good Packaging Yearbook* (July 1971).
8. Commercial Solvents Corp., "Using 2-Nitropropane in Paints Formulated for Application by Electrostatic Spraying," *Tech. Bull.* (1971) **33**.
9. *J. Paint Technol.* (1970) **42**, 646.

RECEIVED March 14, 1972.

9

Solvent Systems for Hydrocarbon Resins

PAUL O. POWERS[1]

Pennsylvania Industrial Chemical Corp., Clairton, Pa. 15025

> *In many applications of low molecular weight hydrocarbon resins, including flooring, adhesives, rubber compounds, inks, and coatings, the best performance is often associated with plasticizers that are marginal solvents rather than perfect ones. The difference between the resin parameter and the plasticizer parameter indicates the place of the system in the Flory-Huggins phase diagram. The separation of phases is responsible for the improved physical properties. While the difference of the parameters readily explains the behavior, the parameters for many industrial materials are not sufficiently well defined, and specific solubility tests must be used to control both resin and plasticizer.*

This study was made to reconcile the behavior of low molecular weight hydrocarbon resins and the behavior of their plasticizers with the solubility parameter and with the Flory-Huggins treatment of phase separation from polymer solutions. These resins are widely used industrially for coatings, floorings, adhesives, rubber compounds, and many other applications. Since they are usually hard and brittle, they are used with rubber, drying oils, plastics, or with plasticizers.

These resins are made by the polymerization of linear and cyclic olefins and diolefins and aromatic olefins to form hard resins averaging from 700 to 1400 in average molecular weight with some over 3,000 molecular weight. The distribution of these polymers has been described (*1*, *2*), and while the distribution is narrow, two or even three peaks are found in the distribution curve. This may be attributable to both the diversity of the feed and to the conditions of polymerization; thus fractions may vary in composition as well as molecular weight.

---

[1] Present address: 742 Ayres Ave., North Plainfield, N. J. 07063.

While the parameter of the solvents is useful in formulating these resins, the solubility parameter is useful in choosing a plasticizer. There must be an appreciable difference between the solubility parameter of the resin and the parameter of the plasticizer to obtain the optimum properties in the compounds in which they are used.

The Flory-Huggins analysis of phase separation in the low molecular weight range is in good agreement with the observed behavior of these resins with plasticizers and solvents. The results indicate that neither a very good solvent nor a very bad one gives satisfactory results with these hydrocarbon resins. The area of optimum behavior may be defined by the difference between the parameter of the resin and the parameter of the solvent or plasticizer.

Solubility Parameter

Hildebrand's solubility parameter (3) is the most useful index of the solubility of resins in solvents and plasticizers. With volatile materials it is derived from the change of vapor pressure with changes in temperature and can be measured very accurately. The parameter of the polymer is the same as that of its monomer. For many hydrocarbon resins, monomers may enter into the polymer, and the complete composition of the resin is not known. In such cases the solubility parameter is estimated from the range of the parameters of the solvents for the polymer.

Parameters of Solvents. The parameters of typical compounds in hydrocarbon solvents are shown in Table I. These values have been derived from the vapor pressure of the pure materials (4).

Styrene and *d*-limonene are included in the table since they are monomers for some resins. Tetralin is used as a model for indene for which no parameter values are available.

Neopentane represents a high concentration of methyl groups. Toluene and trimethylpentane (isooctane) show the effect of the methyl group in reducing the parameter as compared with benzene and *n*-octane.

A typical parameter of the methylene group is seen in octadecane and cyclohexane. The simplest aromatic compound is benzene, and the parameter of nine is in the range of aromatics.

Liquid olefin polymers and mineral oils are often used as plasticizers for the low molecular weight hydrocarbon resins. A rough estimate of the parameters of these materials can be made from their structure, but only approximate values have been offered.

Parameters of Resins. Some of the manufacturers of the low molecular weight hydrocarbon resins have offered estimates (5, 6) of solubility parameters of some of their resins. These values have been established by determining the solvents (in one case at 50% solids) for the

Table I. Solubility Parameter of Hydrocarbons at 25°C

Hydrocarbon	Parameter
Neopentane	6.29
Isoprene	7.4
Diisobutylene	7.7
Octane	7.52
Isooctane	6.86
Octadecane	8.04
Cyclohexane	8.20
Methylcyclohexane	8.13
d-Limonene	8.2
Tetralin	9.5
Benzene	9.16
Toluene	8.8
Styrene	9.3

Table II. Solubility Parameter of Polymers at 25°C

Low Polymers of:	δ	High Polymers of:	δ
Indene	9.5	Ethylene	8.1
Styrene	9.2	Butadiene	8.3
α-Methylstyrene	9.0	Isobutyleme	7.7
Olefines	8.3	Styrene	9.1
Terpenes	8.2	Natural rubber	8.2
Cyclopentadiene	9.1		

resin at room temperature. The solvents may not all be the same if measured at low solids. However the parameters of several polymers determined by precipitation from dilute solution (7) agree with the values determined by other methods.

The solubility parameter will vary with the polarity of the solvent used. Manufacturers report the values for nonbonded, moderately, and highly hydrogen-bonded solvents. In this study parameter values in polar solvents have not been considered since the interpretation of the values with polar compounds is not the same as with hydrocarbons (8).

The values in Table II for low polymers are from the trade literature; values for the high polymers are from the literature (8) or are calculated (9).

While the solubility varies greatly with molecular weight, the solubility parameters remain the same. The parameter for polystyrene remains the same whether the molecular weight is 1,000 or 200,000; in fact the monomer has substantially the same parameter.

Small's method (9) which was used to determine the parameters of some of the high polymers in Table II relies on the structure of the polymer and affords an estimate although it's probably no better than using the parameter of the monomer.

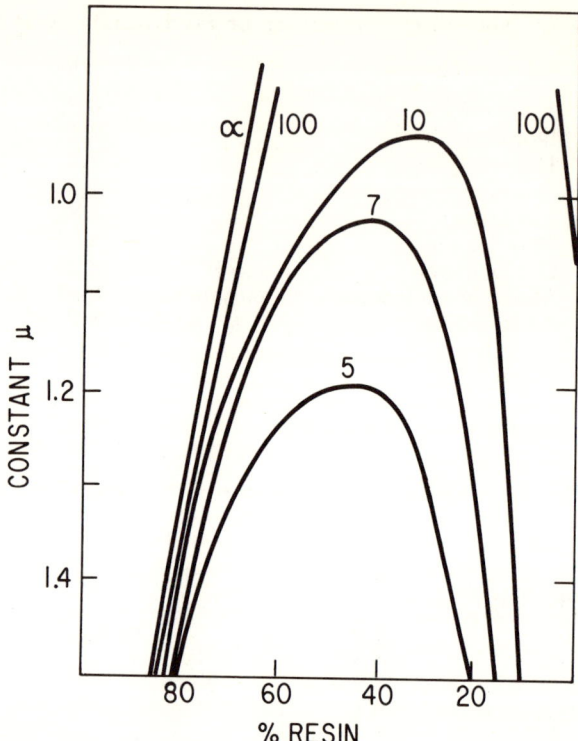

Figure 1. Flory-Huggins phase diagram—low polymers

The low molecular weight hydrocarbon resins have solubility parameter values in the range of 8.2 to 9.5. This might seem narrow, but the solubility behavior of the various resins is quite different and parameter values to the second decimal point are required in choosing a formulation. Usually a solubility test in specific solvents and a cloud point determination are needed for precise control.

Phase Separation

Flory's approach (10) has often been used to study phase separation, but this method does not always predict the exact conditions of separation. The equation of Huggins was used to draw the phase diagram of Figure 1 since values could be calculated over the whole range of concentrations. Also the μ values agreed more closely with the parameter values. This equation is (11):

$$\ln a_1 = \ln V_1 + (1 - X_1/X_2) V_2 + \mu V_2^2$$

where α_1 is the activity of the solvent, V_1, V_2 are the partial volumes of the solvent and polymer, X_1, X_2 are the molar volumes of the solvent and polymer, and μ is the interaction constant.

The interaction constant or μ is proportional to the difference between the solvent parameter and the polymer parameter. Since the polymer and the solvent are about the same size, X_1/X_2 is the inverse of the degree of polymerization. The activity of the solvent has been taken as 1. Figure 1 is the plot of this equation for low molecular weight polymers since the average degree of polymerization of the hydrocarbon resins is about 10. Part of the diagram for high polymers is also shown. The numbers on the curves indicate the degree of polymerization. With high polymers the plasticizer-rich phase contains very little polymer while with low polymers considerable polymer may be present in this phase.

With high polymers, phase separation may occur when μ is greater than 0.5, but when the degree of polymerization is low, as when the degree is 10 (Figure 1), precipitation will not occur until the value is greater than 0.9. In the low range the phase transition is often quite sharp, and the difference in behavior between each successive polymer band can be measured. With high polymers the phase transition is often less clear and may extend over a wider range.

Indene Polymers. Since the commercial resins contain a high content of indene, a commercial coumarone indene resin, Piccoumaron 450-L,

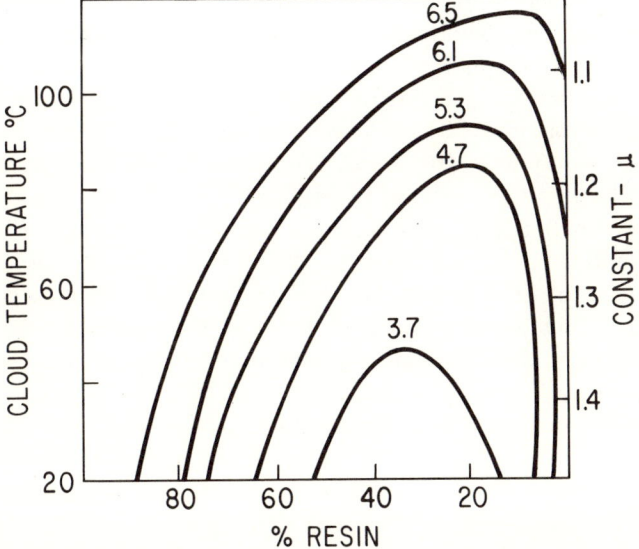

Figure 2. Indene polymer fractions in shell sol 71

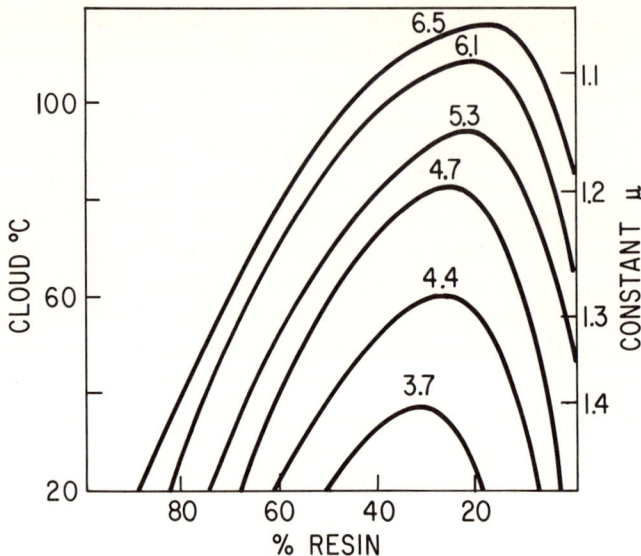

Figure 3. Indene polymer fractions in butanol

was fractionated, and the phase separation of the fractions was studied. Since the resin has a parameter value of 9.5, solvents with two parameter units higher and two units lower were used for the study. For the low parameter solvent Shell-Sol 71, an odorless mineralspirits with a 7.5 parameter value was used. Since no hydrocarbons with an 11.5 parameter value were available, 1-butanol (11.4) was used as the reference solvent with the indene polymer fractions.

Separation was carried out by dissolving the resin in boiling butanol and collecting fractions (2) on cooling at 82°, 75°, 63°, 40°, and 10°C. The resin in solution at 10°C was recovered. Molecular weights were determined by the depression of the melting point of benzene. The phase diagram for Shell Sol 71 and butanol with the fractions are shown in Figures 2 and 3. The numbers on the curves indicate the degree of polymerization of the fraction, the molecular weight divided by 116.

The μ values were calculated for the two solvents by Huggins' method and gave the following equations:

$$\mu = -0.22 + 500/T \text{ for the Shell Sol 71-polyindene}$$

$$\mu = -0.12 + 464/T \text{ for the butanol-polyindene system}$$

where T is the temperature in degrees absolute.

The agreement between Figures 2 and 3 is remarkable as is the general agreement with Figure 1. Reasonable agreement was expected, but such close agreement would not be expected without more exact parameter values for the solvents particularly for the mineral spirits which is a complex mixture.

Applications

When the concept of limited compatibility was first suggested to explain the behavior of certain hydrocarbon resins, there was a general reluctance to accept it; 30 years later there is less than general acceptance. The idea that the best plasticizer gives the best results persists. The solubility parameter indicates the merits of a plasticizer. In all applications of hydrocarbon resins the behavior of the system is measured by the difference of the solubility parameters of the resin and of the plasticizer.

Flooring. Indene polymers were the first resins used in asbestos filled floor tile, and they continue to be used although they have been extended by petroleum resins and reinforced with polystyrene. The tile binder consists of 70% indene resin and 30% mineral oil plus a gelled oil. The filler is a mixture of limestone and asbestos. Aromatic oils give soft sticky mixes, paraffinic oils gave dry, crumbly mixes, and neither made satisfactory tiles.

An important property of flooring is its resistance to indentation which is measured by the rate of indentation of a loaded steel ball. The rate is expressed by the M value in

$$\text{Indentation} = kAt^M$$

where t is the time in minutes and A the indentation at 1 minute. An M of 0.2 is excessive, 0.15 is acceptable, and 0.10 was unattainable since such tiles were very difficult to sheet and exuded plasticizer on aging. When the clouding temperature of the tile binder was plotted against the M value (rate of penetration), a flat plot resulted (Figure 4). (*12*).

In Figure 4 a cloud temperature of 50°C is associated with the desired 0.15 rate of penetration. Figure 2 shows that at 70% resin, a 50°C cloud appears at a degree of polymerization of 6.1 and a μ of 1.3. Indene polymers have a parameter value of 9.5; thus the plasticizer should have parameter of 8.2. which is in the range of naphthenic oils. Also the degree of polymerization is slightly higher than that of indene polymers at 100°C softening point, so the resins are reinforced with some high polymer, often polystyrene.

To meet the specification for the rate of indentation, the tile binder must be in the incompatible range, and this is true whether the resin is polyindene or poly(vinyl chloride). Many other resins have been used to

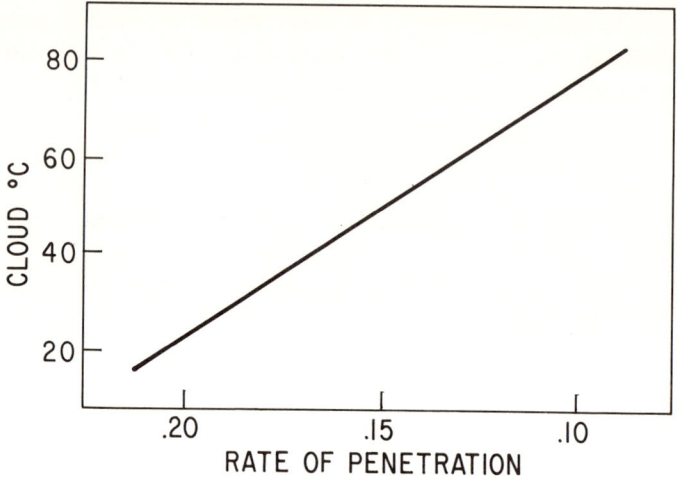

Figure 4. Cloud temperature and rate of penetration

produce tile, but the desired rate of penetration is always associated with this area of limited compatibility.

The interaction between plasticizer and resins must be carefully controlled. Many of the indene resins are controlled within two degrees in the cloud point range. From the relationship

$$\mu = -0.22 + 500/T$$

it is apparent that this is equal to control within 0.005 parameter units, which emphasizes the need of more precise parameter values for industrial plasticizers.

Adhesives. The development of tack is associated with a slight degree of incompatibility which is evidenced only (1) by a loss of gloss. The formation of two phases may in itself be an explanation of the adhesion, one phase being very sticky while the other phase maintains strength.

More recently, somewhat incompatible systems with α-methylstyrene copolymers, ethylene–vinylacetate resins, and paraffin wax (13) have shown exceptional hot tack although the blend is clearly incompatible. Table III shows that the adhesion increases with the cloud point of the blend.

Rubber Compounds. Certain resins have traditionally been termed reinforcing when used with rubber since greater hardness, modulus, and tensile strength are associated with their use (1). These resins are among those less soluble in rubber.

Resins with solubility close to that of rubber (1, 14), the terpene, and olefin polymers have a softening effect on rubber. Indene polymers

Table III. Hot Melt Adhesives

α-Methylstyrene Copolymer, %	Cloud Point, °F	Performance
2		No grab
4	220	No paper break
8	260	25–75% paper break
16	340	75–100% paper break

with more than one parameter unit difference from rubber have pronounced reinforcing effect.

Coatings and Inks. The pronounced effect of certain less soluble resins on the drying rate of paints is associated with the difference between the parameter of the resin and that of the drying oil. However absolute clarity is essential in this field, and close solubility control must be exercised to maintain this clarity.

With printing inks, the parameter of the resin has not been generally used as a measure of performance. However it is apparent that the resins with the best solvent release are those with the greatest spread between the parameter of the resin and the residual solvent. Complete miscibility is regarded as essential, but since drying is often at high temperatures, this requirement is not unduly severe.

Summary

The solubility parameter of resins, solvents, and plasticizers is a valuable index to their behavior in many applications. Coupled with the phase diagram for the resin–plasticizer system, it is possible to determine the area where optimum physical properties are encountered. In this area precise control must be maintained. The parameters of many industrial materials are often not known accurately enough to be useful. However, specific solubility tests are used to give the desired degree of control.

Literature Cited

1. Powers, P. O., *Rubber Chem. Technol.* (1963) **36** (5) 1542.
2. Powers, P. O., "Abstracts of Papers," 158th National Meeting ACS, Sept. 1969, ORPL 47.
3. Hildebrand, J. H., Prausnitz, J. M., Scott, R. L., "Regular Solutions," Van Nostrand Reinhold, New York, 1970.
4. Funk, E. W., Prausnitz, J. M., *Ind. Eng. Chem.* (1970) **62**, 8.
5. "Solubility Parameter of Selected Neville Resins," Neville Chemical Co., Pittsburgh (1966).
6. "Solubility Contours of Hydrocarbon Resins," Pennsylvania Industrial Chemical Corp., Clairton, Pa.
7. Suh, K. W., Corbett, J. M., *J. Appl. Polym. Sci.* (1968) **12**, 2359.

8. Blanks, R., Prausnitz, J. M., *Ind. Eng. Chem. Fundam.* (1964) **3**, 1.
9. Small, P. A., *J. Appl. Chem.* (1953) **3**, 71.
10. Schulz, A. R., Flory, P. J., *J. Amer. Chem. Soc.* (1952) **74**, 4760.
11. Huggins, M. L., "Physical Chemistry of High Polymers," p. 43, John Wiley and Sons, New York, 1958.
12. Powers, P. O., U. S. Patent **2,529,260** (Nov. 7, 1950).
13. Pennsylvania Industrial Chemical Corp., *Bull.* **PPN o34A**.
14. Powers, P. O., *India Rubber World* (1947) **117**, 351.

RECEIVED March 14, 1972.

10

Solvents for Use in Electrodeposition Coatings

C. A. MAY[1]

Shell Development Co., P. O. Box 24225, Oakland, Calif. 94623

> 2-Butoxyethanol, 2-ethoxyethanol, 2-(2-butoxyethoxy)ethanol, isopropyl alcohol, sec-butyl alcohol, methyl isobutyl carbinol, hexylene glycol, methyl isobutyl ketone, and diacetone alcohol were examined as solvents for electrodeposition coating. Three resins were used in the evaluation: a maleinized, epoxy resin-based drying oil ester, an acrylic, and an oleoresinous, trimellitic acid containing alkyd. The glycol ethers were found to be the most useful class of solvent. At least two members of this category could be useful with each vehicle. The ketone was of minor interest. The non-ether alcohols have some merit, and more work appears warranted in this direction. The results suggested that the area of greatest utility may be in solvent blends.

Electrodeposition resins generally consist of classic coating molecules which have been modified by introducing carboxyl groups into their structures. When these carboxyl groups are neutralized with a base, the resin systems are rendered water soluble. The resin solids of electrodeposition systems are normally around 10%, but even at this low solids concentration considerable quantities of resin are involved since the tank volumes may be 60,000 gallons.

One claimed advantage for the electrodeposition process is freedom from organic solvent even though most electrodeposition paints contain small amounts of organic solvents. Although the solvents comprise only a small part (2–3%) of the paint bath, they comprise a relatively large part (ca. 20–30%) of the vehicle. Since it has been forecast that the electrodeposition process will become a major entity in the area of industrial finishing, substantial quantities of organic solvents will be involved.

[1] Present address: 120 Scenic Dr., Orinda, Calif. 94563.

The solvents used for electrodeposition resins perform three functions. First, the solvent facilitates handling during the preparation of the aqueous solution since these carboxyl containing materials are either resinous or sticky semi-solids. Second, the solvent may accompany the vehicle during deposition and therefore can influence the flow characteristics during post-deposition bake or cure. Third, the solvent can confer better water solubility characteristics on the resin and thus help to maintain bath stability, a factor of prime importance in electrocoating.

Despite the importance of solvents in electrodeposition, technical literature on the subject is sparse. To the author's knowledge there have been only two publications on this subject (*1, 2*). There is also a third, unpublished (*3*) effort which indicates that the glycol ethers are a particularly useful class of solvent. However, all of these studies are narrow in scope in that only one electrodeposition resin was used in each case. The present study encompasses three of the more popular resin systems.

The solvents chosen for this study were 2-butoxyethanol, 2-ethoxyethanol, 2-(2-butoxyethoxy)-ethanol, isopropyl alcohol, *sec*-butyl alcohol, methyl isobutyl carbinol, hexylene glycol (2,4-dihydroxy-2-methylpentane), methyl isobutyl ketone, and diacetone alcohol. The structural formulas are shown in Figure 1. Three resins were chosen as representative of a cross-section of the types of vehicles currently being used. The resins were a maleinized, epoxy resin based, drying oil ester (typifying a high quality automotive or appliance primer), an acrylic resin (used primarily for one-coat finishes), and a trimellitic based oleoresinous alkyd (representing a low cost drying oil type vehicle).

Since 27 possible combinations of resin and solvent were possible, each solvent resin system could not be examined over a wide range of concentrations. Accordingly, the investigation was conducted in two parts. In the first part all possible combinations of resin and solvent were evaluated at 70% non-volatile concentration. In the second part the solvents which appeared the most interesting were examined at other concentrations. The glycol ethers generally proved to be the most versatile for electrodeposition. As the studies progressed, however, certain other areas of interest developed which may be worthy of future study.

Resins and Definition of Terms

The terminology is briefly defined for those not familiar with the technique of electrodeposition; further details are given in the experimental section.

Resins. The epoxy ester was prepared from one of the solid grades of epichlorohydrin/bisphenol-A type resin. The resin was first esterified with benzoic acid to 40% of the available esterifiable groups. The re-

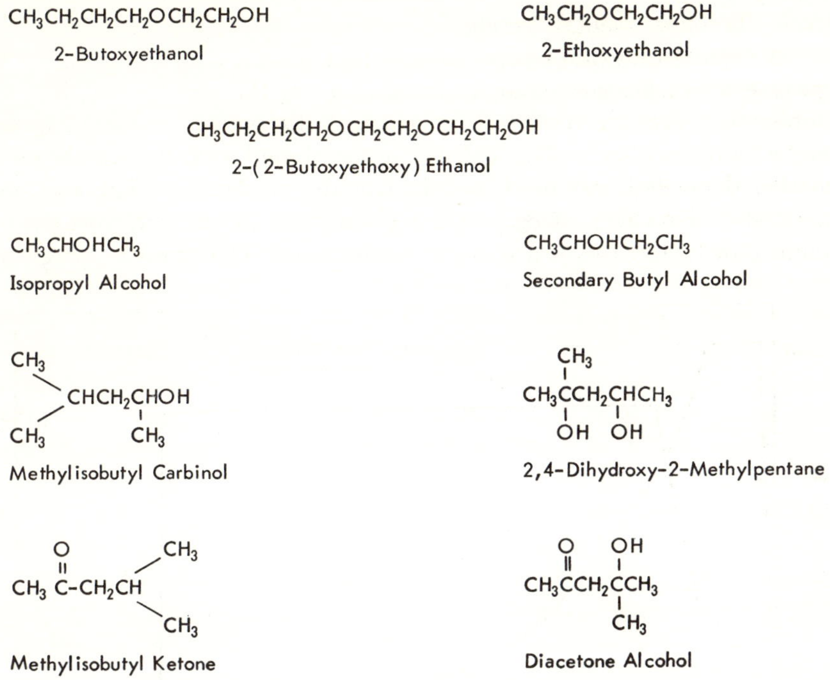

CH₃CH₂CH₂CH₂OCH₂CH₂OH
2-Butoxyethanol

CH₃CH₂OCH₂CH₂OH
2-Ethoxyethanol

CH₃CH₂CH₂CH₂OCH₂CH₂OCH₂CH₂OH
2-(2-Butoxyethoxy) Ethanol

CH₃CHOHCH₃
Isopropyl Alcohol

CH₃CHOHCH₂CH₃
Secondary Butyl Alcohol

(CH₃)₂CHCH₂CHOHCH₃
Methylisobutyl Carbinol

CH₃C(OH)(CH₃)CH₂CH(OH)CH₃
2,4-Dihydroxy-2-Methylpentane

CH₃C(O)CH₂CH(CH₃)₂
Methylisobutyl Ketone

CH₃C(O)CH₂C(OH)(CH₃)CH₃
Diacetone Alcohol

Figure 1. Structural formulas of solvents investigated

maining 60% was then esterified with linseed oil fatty acids giving an essentially hydroxyl-free ester. Water solubility was achieved by subsequent reaction of the ester with maleic anhydride and neutralization with a base (triethylamine).

The acrylic resin used was Rohm and Haas Co.'s QR-496 (4), so the exact nature of the polymer is not known. It is hydroxy functional, thus requiring an aminoplast for proper cure; commercially modified hexamethylol melamine resin was used for this purpose. During the first part of the investigation the acrylic resin was used without the melamine resin, but for final evaluation it was incorporated to give a more realistic commercial coating system.

The oleoresinous, trimellitic anhydride-based alkyd was also obtained commercially; it is a 46% oil length safflower oil resin. In addition to the trimellitic anhydride the resin also contains hydrogenated bisphenol-A and isophthalic acid (5).

Definition of Terms. Only the deposition characteristics, including appearance of the baked coating, were considered in evaluating the solvents. In the readers interest, the important terms used throughout this report (in italics below) are described. Details of the coating process are in the experimental section.

A typical time/amperage/voltage plot is shown in Figure 2 for the epoxy ester in 2-ethoxyethanol. Integration of the area under the time-amperage curve was used as a measure of the power consumed in coulombs (amperes/sec). This calculation combined with the weight of the coating deposited was used to calculate the *coulombic efficiency* (mg/coulomb). However, since power consumption in electrodeposition is a minor cost factor, this is a minor consideration. The time-amperage plot

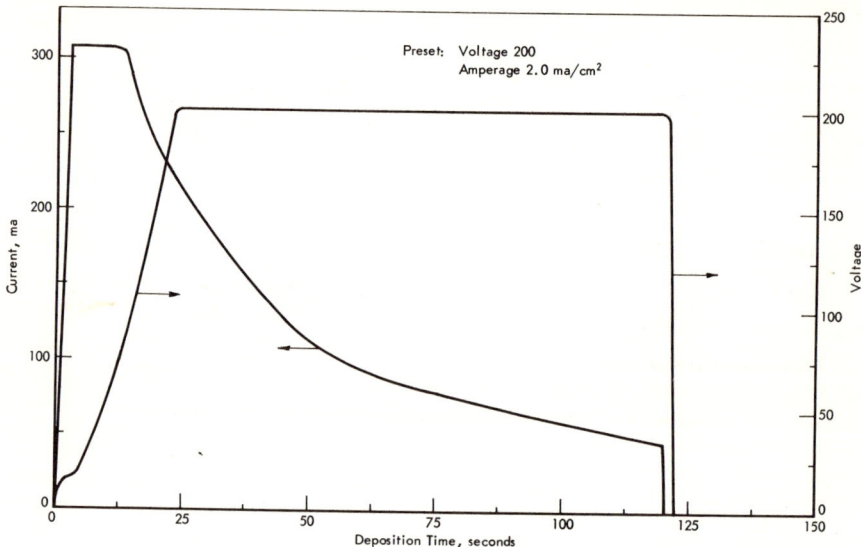

Figure 2. Time–amperage–voltage plot, epoxy ester in 2-ethoxyethanol, 70% non-volatile

is also one method for determining the *rupture voltage*. When the coating ruptures, there is an increase in the time-amperage curve. The rupture voltage was also used to select the *maximum operating voltage* (MOV). This was generally chosen at about 10 volts below the rupture voltage.

From the amperage curve we determined the final amperage, *i.e.*, the amperage at the end of the coating period. A final *wet film resistance* was calculated from Ohm's Law in ohms/cm^2/mil. This measurement is related to *throwing power—i.e.*, the ability of the coating to reach remote areas of the object being painted. This is misleading because it minimizes the importance of throwing power, and actually a number of other factors may be involved. A much better way would be to measure throwing power, perhaps by the Ford pipe test, but time did not permit it during this investigation.

The *film weight* was measured for given conditions and expressed in mg/sq cm. The *film thickness* was determined on the front (side facing the cathode) and back of the panels. The ratio of these thicknesses is a test of uniformity and an indication of throwing power. The film thickness on the front was used to prepare contour map of thickness as a function of voltage and solids concentration for a given resin–solvent com-

Table I. Solution Properties of Vehicles at 30% Non-volatile

Resin system	Solvent	Properties of H$_2$O Solution			
		Viscosity, Brookfield	pH	Conductivity, μmhos	Light Transmission at 5560 A, %
Epoxy ester	2-butoxyethanol	74.4	9.50	1266	<1
	2-ethoxyethanol	50.4	9.60	1190	57
	2-(2-butoxyethoxy) ethanol	122.4	9.50	1266	<1
	isopropyl alcohol	resin only soluble hot			
	sec-butyl alcohol	—	9.70	1075	40
	methyl isobutyl carbinol		9.70	1010	<1
	hexylene glycol	insoluble			
	methyl isobutyl ketone	gelled on addition of triethylamine			
	diacetone alcohol	gelled on addition of triethylamine			
Acrylic	2-butoxyethanol	78.0	8.45	1369	99
	2-ethoxyethanol	34.0	8.51	1266	100
	2-(2-butoxyethoxy) ethanol	163.6	8.48	1370	67
	isopropyl alcohol	33.0	8.45	1266	65
	sec-butyl alcohol	41.0	8.51	1266	100
	methyl isobutyl carbinol	78.4	8.70	1235	96
	hexylene glycol	492.5	8.51	1282	100
	methyl isobutyl ketone	29.6	8.60	1204	82
	diacetone alcohol	114.0	8.51	1266	100
Trimellitic alkyd	2-butoxyethanol	9.4	9.50	1449	91
	2-ethoxyethanol	6.5	9.72	1111	15
	2-(2-butoxyethoxy) ethanol	18.5	9.50	1429	98
	isopropyl alcohol	6.5	9.40	1205	91
	sec-butyl alcohol	7.3	9.30	1282	96
	methyl isobutyl carbinol	10.6	9.40	1176	42
	hexylene glycol	76.8	9.70	1190	92
	methyl isobutyl ketone	aqueous solution would not form			
	diacetone alcohol	10.0	7.35	1613	82

bination. Another consideration in evaluating co-solvents is the general appearance of the film after baking.

Solution stability of the aqueous resin system is also a key factor with regard to co-solvents for electrodeposition paints. It was determined at both room temperature and 40°C. The criteria used were pH, conductivity, and haze. Haze is a function of light transmission at 5560 A and was measured using a Cary model 14 spectrometer. All tests were run in sealed containers.

Preliminary Evaluation

The 27 possible combinations of resin and solvent were studied at one concentration, 70% non-volatile; the most interesting combinations were then evaluated further.

In this first part of this study the properties considered were the aqueous and solvent solution properties, rupture voltage, wet film resistance, film thickness and weight, coulombic efficiency, and surface appearance. The viscosities and the aqueous solution conductivity and haze are summarized in Table I. Because of their solubility characteristics,

Table II. Summary of Preliminary Screening of Solvents

Solvent	Rupture voltage	Film Resistance	
		Peak	General level
2-butoxyethanol–epoxy ester	200	<190	
2-ethoxyethanol	270–275	205	+3
2-(2-butoxyethoxy) ethanol	240–250	135	+2
sec-butyl alcohol	430–440	<400	+2
methyl isobutyl carbinol	360	<350	+3
2-butoxyethanol–Acrylic	80	< 60	
2-ethoxyethanol	170–180	<160	+3
2-(2-butoxyethoxy) ethanol	120–130	<100	+1
isopropyl alcohol	120–130	93	+2
hexylene glycol	160–170	124	+3
diacetone alcohol	170–180	128	+3
2-butoxyethanol–TMA alkyd	80	< 70	
2-ethoxyethanol	220–230	>140	+3
2-(2-butoxyethoxy) ethanol	130–140	<120	+1
isopropyl alcohol	220–240	150	+3
sec-butyl alcohol	170–175	<160	+2
hexylene glycol	200–230	None	+2
diacetone alcohol	235–240	−140	+3

[a] Key: +3 Definitely superior, +2 better, +1 slightly better, −3 definitely inferior, −2 poorer, −1 slightly poorer, and = About the same.

certain combinations were of little interest. The solvents eliminated at this point were isopropyl alcohol, hexylene glycol, and diacetone alcohol with the epoxy ester and methyl isobutyl ketone with both the epoxy ester and the trimellitic alkyd.

Table II summarizes the overall results and indicates which solvents were the best choices for further evaluation. Fortunately, the generally recommended co-solvent for each of the three resin systems was 2-butoxyethanol. Comparisons thus could be made using this solvent as a common base. The data are not quantitative but are expressed on a better-than or poorer-than basis, with $+3$ being superior and -3 being inferior.

The columns labeled S position can best be understood by examining Figure 3 where the wet film resistance is plotted as a function of application voltage for various solvents using the epoxy ester electrocoating system. Each curve is a portion of the letter S. The other two resins also displayed this characteristic. The lower, middle, and upper parts of the letter are readily seen. The best results should be obtained when the curve resembles the upper part of the S, rupture voltage permitting. The lower part of the S is meaningless because wet film weight, thickness, and wet film resistance are low and there is an upper limit where the

for Electrodeposition Resins[a]

Thickness and Weight		S position	Coulombic efficiency	Surface appearance
General level	Voltage response			
0	+2	0	−1	−1
+1	+2	0	−1	0
+1	+1	+1	−2	−2
+2	+2	+2	+2	−1
+1	0	+3	+2	0
0	+2	0	0	0
0	−3	+2	+2	0
+2	+1	+3	+2	0
+2	+1	+3	+2	+1[b]
+3	+3	+1	+3	+1
+2	+2	+1	0	+2
+3	+3	+1	+2	+1
+3	+3	+1	+1	+1
+3	+3	+1	+1	+2
+3	+3	+1	+3	+1

[b] Diacetone alcohol seemed partially to cure the resin even in the absence of a melamine resin.

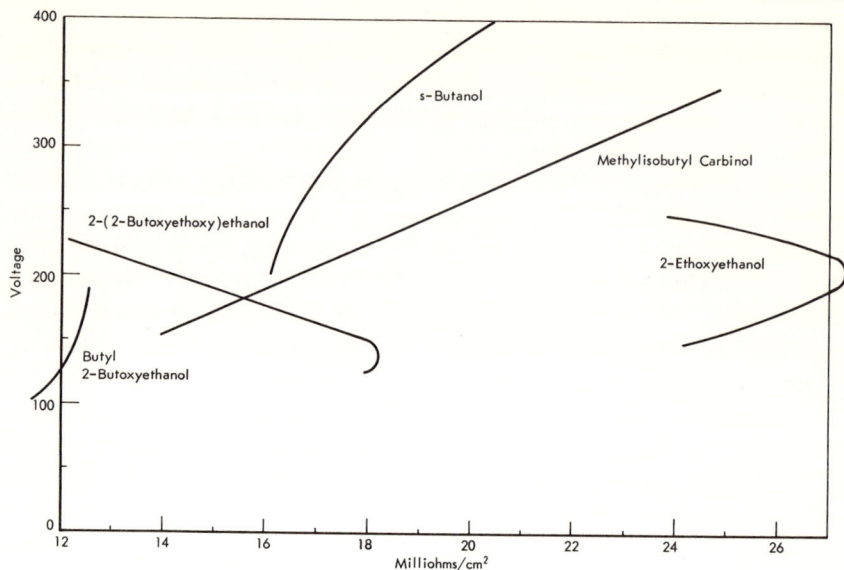

Figure 3. Effect of cosolvent on wet film resistance of the epoxy ester

voltage has little effect. The center portion of the S properties would tend to decrease with increasing voltage. This would probably limit the throwing power since even though the voltage increases, this gain may be offset by a decline in overall film resistance. This observation can be a useful tool for other investigators.

The results on storage tests of the aqueous solution at room temperature and 40°C showed relatively small changes and hence were not a factor in the preliminary considerations. Analysis of the preliminary data for each resin is as follows.

Epoxy Resin Ester. Isopropyl alcohol, hexylene, glycol, methyl isobutyl ketone, and diacetone alcohol were ruled out on solution properties. 2-Ethoxyethanol, *sec*-butyl alcohol, and methyl isobutyl carbinol gave the highest rupture voltages and greater control over film thickness by voltage change than by 2-butoxyethanol. Despite these advantages, the appearance of the films would probably be commercially unacceptable pending pigmentation studies. Thus, 2-(2-butoxyethoxy)ethanol was the only remaining candidate. Because the glycol ethers, as a class, gave interesting results, 2-methoxyethanol, 2-(2-methoxyethoxy)ethanol, and 2-(2-ethoxyethoxy)ethanol were also considered in the more detailed portion of the investigation.

Acrylic Resin. The acrylic resin had excellent solubility characteristics in all nine solvents. On the basis of electrocoating experiments, *sec*-butyl alcohol, methyl isobutyl carbinol, and methyl isobutyl ketone were ruled out. It is not possible to coat these systems above 30–40 volts. All of the remaining candidate solvents had rupture voltages and peak film resistant voltages greater than the 2-butoxyethanol control. 2-Ethoxyethanol and 2-(2-butoxyethoxy)ethanol were selected because of the general utility of glycol ethers in electrodeposition. The other solvent chosen for further study was diacetone alcohol which appeared to contribute to the cure of the resin system.

Trimellitic Alkyd. Methyl isobutyl ketone was ruled out for poor aqueous solubility characteristics. 2-Ethoxyethanol and 2-(2-butoxyethoxy)ethanol were chosen in keeping with the selection for the other two

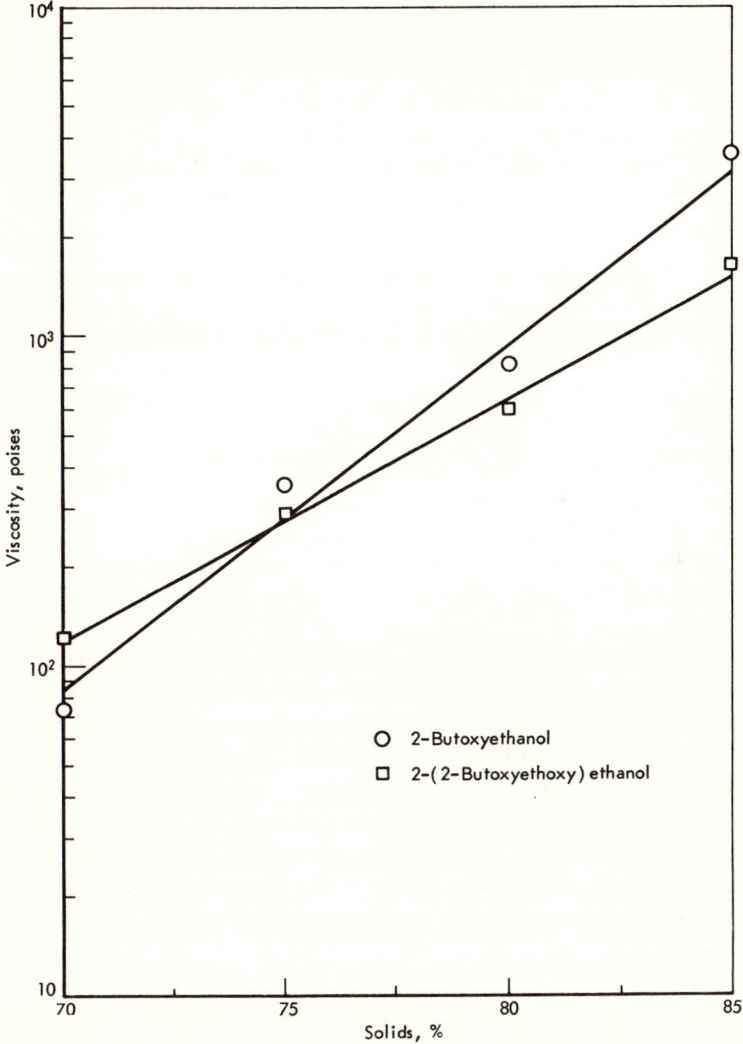

Figure 4. Dilution curves for the epoxy ester

resins. The fourth solvent, in addition to the 2-butoxyethanol control, was hexylene glycol. Despite its high viscosity, it offered numerous advantages over 2-butoxyethanol and gave the most attractive film.

Detailed Studies on the Better Candidates

Additional study was done on the best candidates. The results are discussed below for each resin, and the advantages and disadvantages of the various solvents are noted. With these data the user of an individual class of resins should be better able to choose a starting point based on his process requirements.

Many factors enter into this decision. For instance, the viscosity of the solvent solution prior to let down could be significant if the object to be coated is relatively simple and high throwing power is not required. Conversely, high throwing power may be needed for coating a complex shape making solvent concentration and type important.

Table III. Rupture Voltages of Epoxy Ester Solutions

Solvent	% Resin Solids			
	70	75	80	85
2-butoxyethanol	200	310	385	440
2-methoxyethanol	380	—	—	560
2-ethoxyethanol	270–75	—	—	—
2-(2-butoxyethoxy) ethanol	240–50	300	285	285
2-(2-methoxyethoxy) ethanol	290	—	—	480
2-(2-ethoxyethoxy) ethanol	320	—	—	510
methyl isobutyl carbinol	360	—	460	460
sec-butyl alcohol	430–40	—	—	560

Solvents for the Epoxy Ester. The viscosity of ester solutions of the two candidate solvents as a function of percent solids is shown in Figure 4. There is little difference between the two. Concentrations in the range of 70–80% would probably give the optimum performance in an electrodeposition paint system.

The effect of solvent concentration and type on rupture voltage appears in Table III. At the low solids concentration 2-butoxyethanol gives the poorest results. However, as the solid concentration increases, the rupture voltage also increases dramatically compared with the other solvents examined. 2-(2-Butoxyethoxy)ethanol, is relatively unaffected by concentration.

For the reference standard, 2-butoxyethanol, the effect of application voltage and solids concentration on film thickness is shown in Figure 5. Only a narrow range of thicknesses can be obtained regardless of the applied voltage. Subsequent data indicate that this may be a characteristic of the maleinized epoxy ester.

As Figure 6 shows, the wet film resistance improves with increasing voltage at the higher solids concentrations. Thus, throwing power should also increase. At the lower concentration voltage appears to have little effect.

Figure 7 shows that the maximum coulombic efficiency with both solvents is about 80% solids. Compared with other parameters, coulombic efficiency varied only slightly with solvent concentration and was relatively unaffected by operating voltage.

Uniform film thicknesses were obtained on the fronts and backs of the panels. An increase in solvent concentration and a decrease in application voltage led to some degradation of film appearance. Both variables influenced film thickness, and thinner-electrodeposited films tend to be less attractive.

The only solvent of the nine candidates which compared favorably with 2-butoxyethanol was 2-(2-butoxyethoxy)ethanol. As pointed out in Figure 4, solution viscosity was not a major factor. It was noted from Table 3, however, that this solvent gives a higher rupture voltage at lower solvent concentrations. This could be important if a compromise were needed between throwing power and the handling or let down requirements of the resin solution for the plant. Figure 8 shows that this solvent

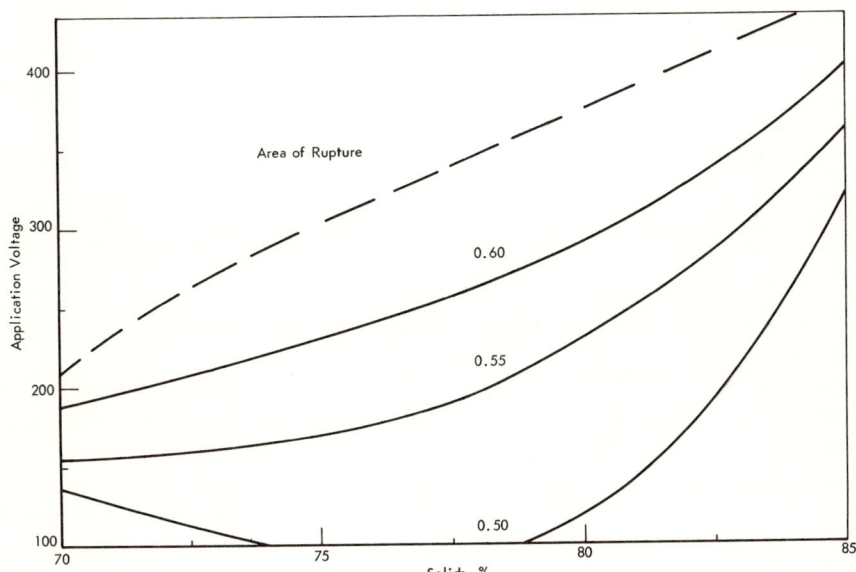

Figure 5. *Epoxy ester–2-butoxyethanol: effect of voltage and solids on film thickness (mils)*

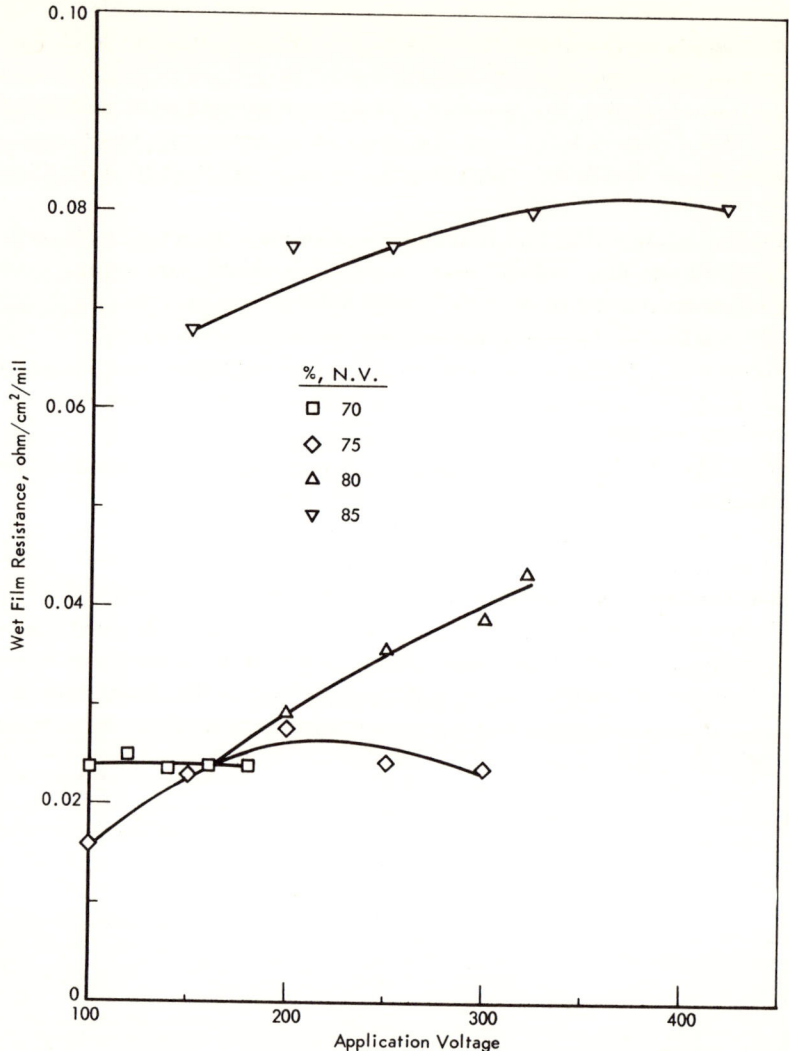

Figure 6. Epoxy ester–2-butoxyethanol: voltage vs. wet film resistance

also does not afford a wide range of film thicknesses. Considering the wet film resistance in Figure 9, 2-(2-butoxyethoxy)ethanol offers marginal advantages in the mid-concentration ranges (75 and 80%). These data also show that the wet film resistance approaches a maximum at voltages below the maximum operating voltage. This may indicate some disadvantage in the throwing power, but such a conclusion would only be justified after additional study.

The coulombic efficiency is not a basis for discriminating between two solvents. As with the control, this varied only slightly with voltage at a given solids concentration for 2-(2-butoxyethoxy)ethanol. Film uniformity over the entire surface should be of little concern. Over this concentration range 2-(2-butoxyethoxy)ethanol gave better-appearing films, but these systems were more sensitive to application voltage. Since this could result from film thickness variations, the point should be carefully checked with individual paints.

2-(2-Butoxyethoxy)ethanol should be considered a useful solvent in epoxy ester based electrodeposition coatings. Although the differences between the solvents were small, there were certain advantages compared

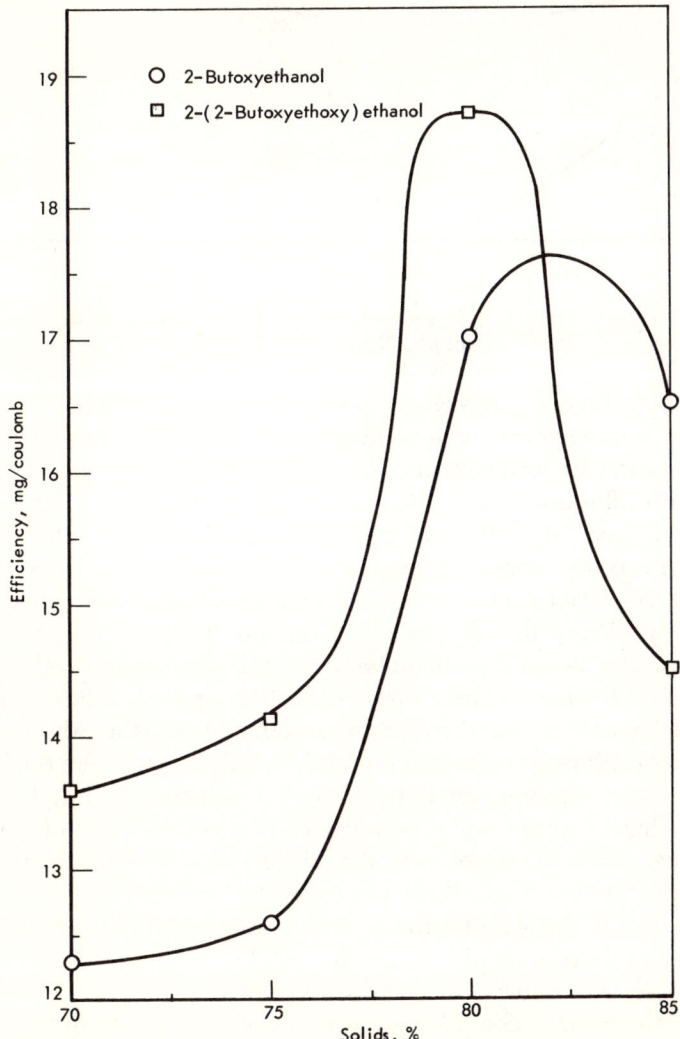

Figure 7. Coulombic efficiencies for the epoxy ester

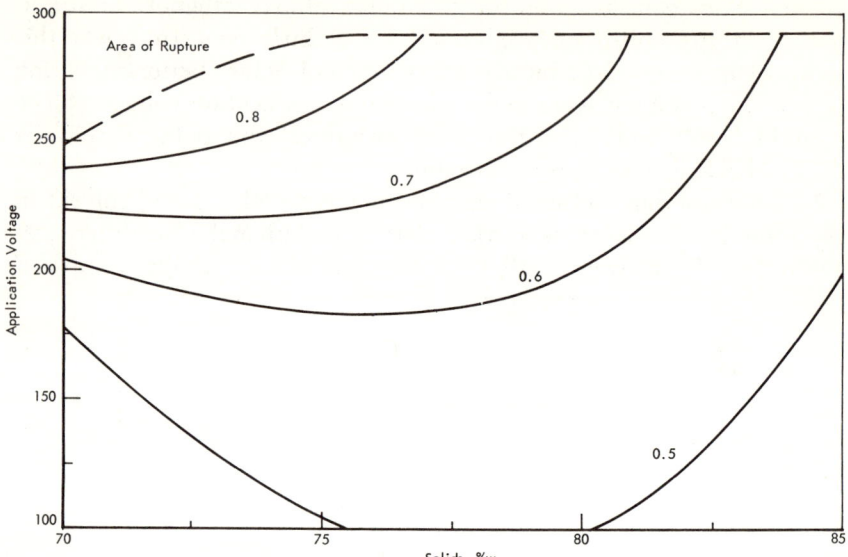

Figure 8. Epoxy ester–2-(2-butoxyethoxy)ethanol: effect of voltage and solids on film thickness (mils)

with 2-butoxyethanol: (a) slightly lower viscosities at high concentrations, (b) some improvement in film appearance, and (c) probably better throwing power at low voltages.

Because film appearance limited the choice of solvents to only two from the original list, data were generated on other glycol ethers. 2-Ethoxyethanol was not promising because aqueous electrodeposition solutions could not be formed at non-volatile concentration greater than 70%. 2-(2-Ethoxyethoxy)ethanol was ruled out on the basis of general film appearance, but 2-methoxyethanol and 2-(2-methoxyethoxy)ethanol may be worthy of further investigation. Viscosities at high solids concentrations were lower than found with 2-butoxyethanol. Further, these solvents gave higher rupture voltages and a slightly wider range of film thicknesses.

Two of the alcohols, *sec*-butyl alcohol and methyl isobutyl carbinol, gave very high rupture voltages and wet film resistances but poor film appearance. Thus, these solvents should be considered in mixtures, and further investigation along these lines may be warranted.

Solvents for the Acrylic Resin. The choice of the best candidate in this case was more difficult. Since the product had excellent solubility characteristics, a number of choices were possible. Again using 2-butoxyethanol as the control, comparisons were made with results obtained from solutions in 2-(2-butoxyethoxy)ethanol, 2-ethoxyethanol, and diacetone alcohol.

The effect of solvent concentration on viscosity is shown in Figure 10. There is little to choose from among the various solvents. Generally higher viscosities were obtained with the 2-(2-butoxyethoxy)ethanol, but the differences are small.

Table IV shows the rupture voltages of the various solvents at different solids concentrations. At all concentrations 2-ethoxyethanol is clearly superior, but diacetone alcohol shows some advantages at the lower solids concentrations. Figure 11 illustrates the effect of voltage and solids content on film thickness. This property did not vary markedly with any of the solvents evaluated. In all cases the contour lines had a slope which approximates the curve of rupture voltages *vs.* non-volatile concentration. Figures 12 and 13 show the wet film resistances for 2-butoxyethanol and

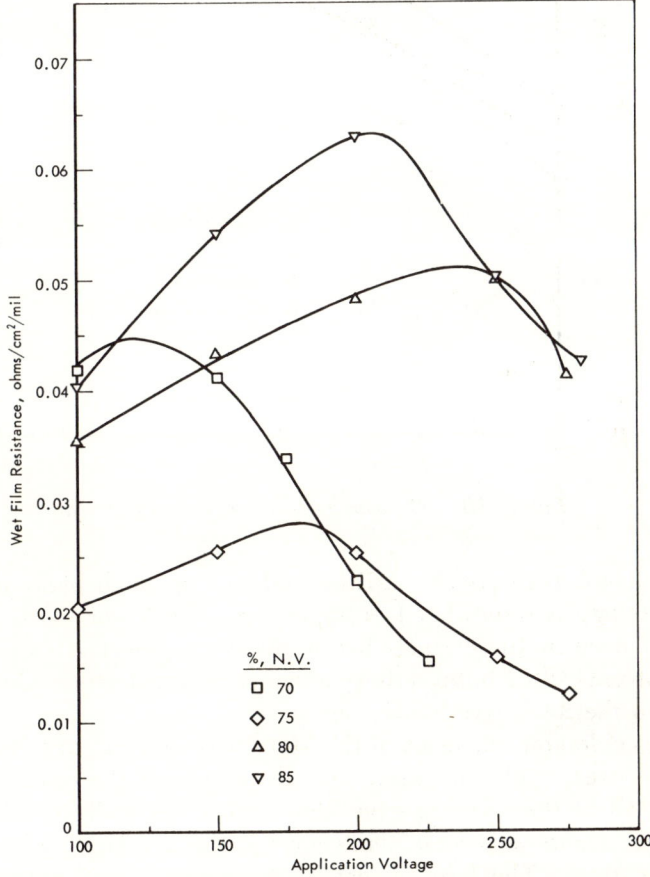

Figure 9. Epoxy ester–2-(2-butoxyethoxy)ethanol: voltage vs. wet film resistance

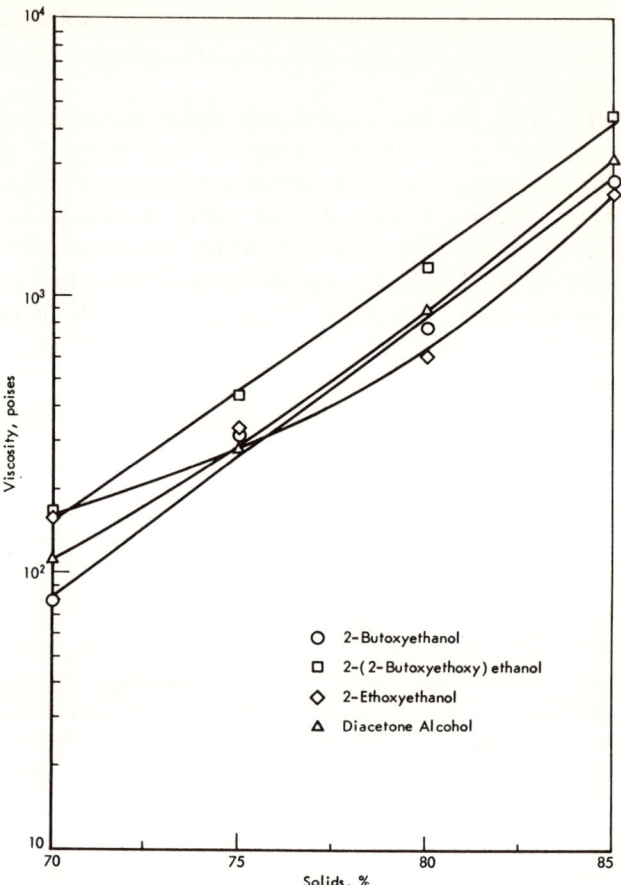

Figure 10. Dilution curves for the acrylic resin

2-ethylethanol, respectively. 2-Ethoxyethanol again displays advantages; the advantage is somewhat less apparent at the lower solids concentration, but even at these levels better throwing power is expected. The other solvents, 2-(2-butoxyethoxy)ethanol and diacetone alcohol, were similar to the 2-butoxyethanol control.

The coulombic efficiency at the MOV was too close to be meaningful. All values were high compared with the values of the other two resins. Further, all of the solvents gave films with a generally satisfactory appearance. There was some tendency for the acrylic to pin hole with melamine resin. The best solvent in this regard again appeared to be 2-ethoxyethanol. Storage stability of the aqueous solutions, regardless of the solvent, presented no problems.

2-Ethoxyethanol is an interesting solvent to use with the acrylic resin. Its main strength is that it gives the highest rupture voltages and best wet film resistance of the solvents evaluated. In addition, the viscosity of the resin-solvent solution is among the lowest in this series. Films can be applied over a wide range of thicknesses, and as judged from the wet film resistance and high rupture voltages, the throwing power should be better than with 2-butoxyethanol. The one disturbing

Table IV. Rupture Voltages of Acrylic Resin Solutions

Solvent	% Resin Solids				
	70[a]	70	75	80	85
2-ethoxyethanol	170–80	235	260–70	260–75	330–40
2-butoxyethanol	80	160	190	220	240
2-(2-butoxyethoxy)ethanol	120–30	180	215	250	240
isopropyl alcohol	120–30	—	—	—	—
sec-butyl alcohol	40–50	—	—	—	—
hexylene glycol	160–70	—	—	—	—
diacetone alcohol	170–80	240	220	220	240

[a] Values obtained in preliminary studies without added melamine resin.

factor is that at the higher application voltages, wet film resistance has a pronounced tendency to decline. This, however, appears to be a characteristic of the electrodeposition resin and has little to do with the solvent. This effect was noted with all four solvents and is also seen with other resins. Diacetone alcohol may be of some interest but it is not as good as 2-ethoxyethanol. It would be interesting to see whether the enhanced cure observed in the preliminary work could be detected in the films cured with the melamine resin.

Solvents for the Trimellitic Alkyd. The four solvents evaluated with this resin were the 2-butoxyethanol control, 2-ethoxyethanol, 2-(2-butoxyethoxy)ethanol, and hexylene glycol. This alkyd displayed a pronounced tendency to sag. It is suspected that the viscosity of the neat resin is somewhat lower than desired from the standpoint of film forming characteristics. By contrast the wet film resistances observed were an order of magnitude higher than those obtained with either of the other two resins. It would be interesting to know what effect this has on throwing power.

From the standpoint of viscosity (Figure 14), 2-ethoxyethanol is the most effective solvent. Even at a non-volatile concentration of 85% the solution would be acceptable for most commercial installations.

The rupture voltages for the various resin–solvent combinations are shown in Table V. Of the candidates under consideration 2-ethoxyethanol and hexylene glycol gave the highest values. This is also reflected somewhat in the plots of film thickness as a function of voltage and

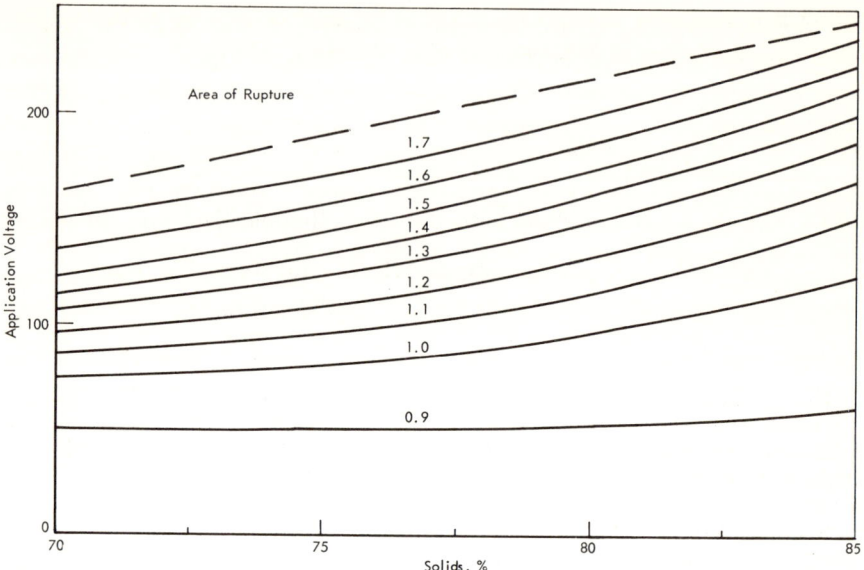

Figure 11. Acrylic–2-butoxyethanol: effect of voltage and solids on film thickness (mils)

percent solids. Illustrative of this point are Figures 15 and 16, where 2-butoxyethanol is compared with 2-ethoxyethanol. Control of the coating operation in the area where greater film thickness can be attained with 2-ethoxyethanol may be difficult. Further, in the case of 2-ethoxyethanol above 75% solids it is difficult to achieve films with a thickness greater than about 0.7 mil without rupture.

There is little difference among the various solvents in coulombic efficiency at the MOV. They all seemed satisfactory for electrocoating operations. Comparing Figures 17 and 18 we see that ethyl 2-ethoxyethanol (Figure 18) gives the best wet film resistance. Hexylene glycol was similar to 2-ethoxyethanol, and 2-(2-butoxyethoxy)ethanol was similar to the 2-butoxyethanol control. As with the other resins, wet film resistance decreases with increasing solvent concentration.

The storage stability of all of the aqueous solutions appeared to be satisfactory. However 2-(2-butoxyethoxy)ethanol caused the film to sag during baking. Accordingly, this solvent is not recommended unless the viscosity of the base resin can be increased. Overall, 2-ethoxyethanol and hexylene glycol are interesting and useful solvents for the trimellitic alkyd. Compared with 2-butoxyethanol they seem to possess greater throwing power by virtue of the higher rupture voltage and greater wet film resistance. In addition, marginal advantages were noted in film appearance. The major disadvantage was that of obtaining sufficiently thick films; however, it may be possible to control this by reducing the solids concentration at some sacrifice in throwing power.

Experimental

Preparative Details, Maleinized Epoxy Ester. The reaction was carried out in a 5-liter, four-neck kettle equipped with a stirrer, nitrogen bubbler, thermowell, and a Dean-Stark trap surmounted by a condenser. The entire reaction was run under nitrogen in five basic steps.

STEP I. PREPARATION OF PROPER MOLECULAR WEIGHT RESIN. The charge was Epon 829, precatalyzed epoxy liquid resin (968.0 grams) and bisphenol-A (259.0 grams). These components were heated to 138°C (280°F), the heat removed, and the exotherm rose to 176°C (349°F) in 10 minutes. The charge was held at this temperature for ½ hour.

STEP II. PREPARATION OF X-BE-4 ESTER. To the above was added a charge consisting of: benzoic acid (372 grams), benzyldimethylamine (0.09 phr on resin) (1.40 grams), and xylene (3% on charge) (49.0 grams).

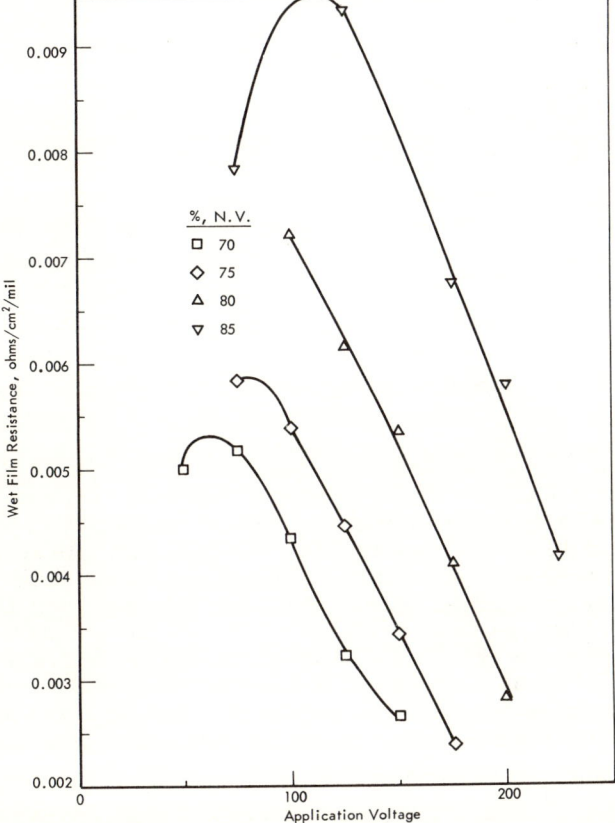

Figure 12. Acrylic–2-butoxyethanol: voltage vs. wet film resistance

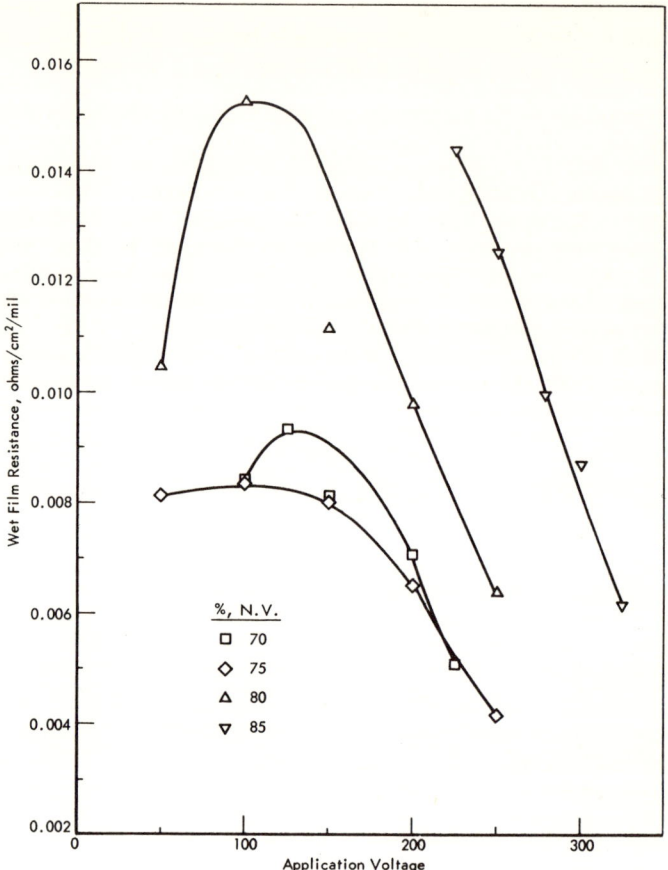

Figure 13. Acrylic–2-ethoxyethanol: voltage vs. wet film resistance

The amine was added first, then the acid and the xylene. The kettle was heated to 190°C (374°F) in 10 minutes and held for an acid number of 18.1 mg KOH/gram (25 minutes). We then added 0.37 gram stannous octoate and heated the mixture to 240°C (464°F). An acid number of 0.84 was reached after 3½ hours. The xylene concentration was adjusted to maintain a gentle reflux during this step and Step III below.

STEP III. PREPARATION OF X-BE-4 L-6 ESTER. To the above mixture we added 1280 grams of linseed oil fatty acids, and the charge was heated to 270°C (518°F) in 1½ hours. The charge was cooled overnight and reheated to reaction temperature the following morning. It was held at this temperature for 6¾ hours, giving an acid value of 11.6. During the last hour the gas sparge rate was increased, and the xylene was removed from the charge.

STEP IV. MALEINIZATION. The charge was cooled to 200°C (392°F), and 152 grams of maleic anhydride were added. Just prior to adding the anhydride the amount was calculated from the relationship:

$$\frac{60 \times \text{wt of ester}}{1144 - 60}$$

to give a product with a theoretical acid value of 60.0 assuming that all carboxyls are available and none are half ester. The reaction temperature was held at 200°C for 2 hours. The final acid value (benzene–ethanol) was 42.3, non-volatile 96.0%.

STEP V. SOLVENT LET-DOWN. For most of the work done in this investigation the product was cooled in the neat form so that various solvents could be used. The normal let-down procedure involved cooling the resin to 132°C (270°F) and adding a sufficient amount of 2-butoxyethanol to make an 80% non-volatile solution.

Typical Aqueous Solution Preparation (70% non-volatile in 2-Butoxyethanol). A cold cut was made of 105 grams of the epoxy ester and 45 grams of 2-butoxyethanol. To this solution were added 8.41 grams triethylamine, 100% neutralization based on the pretitrated acidity. Wa-

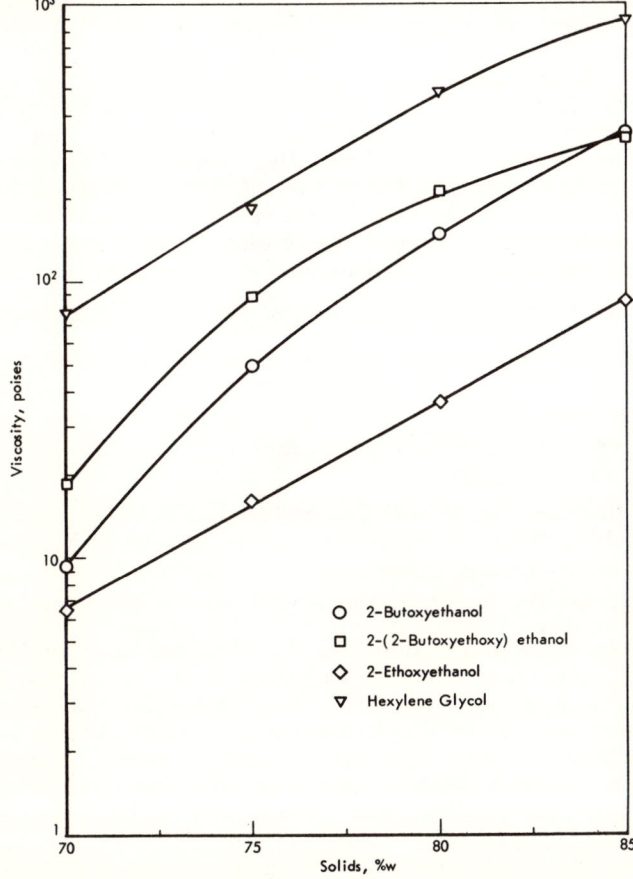

Figure 14. Dilution curves for the trimellitic alkyd

Table V. Rupture Voltages of Trimellitic Alkyd Solutions

Solvent	% Resin Solids			
	70	75	80	85
2-ethoxyethanol	220–30	270–75	325	320–30
2-butoxyethanol	80	130	170	230
2-(2-butoxyethoxy) ethanol	120	150	200	220
isopropyl alcohol	220–40	—	—	—
sec-butyl alcohol	170–75	—	—	—
methyl isobutyl carbinol	50	—	—	—
hexylene glycol	200–30	320	320	390
diacetone alcohol	235–40	—	—	—

ter (870 ml) was then added from a dropping funnel with vigorous stirring to give a 10% resin solids solution neglecting the added amine. During dilution the viscosity increased to ca. 30–40% solids, then dropped rapidly to a watery consistency. The electrodeposited films were baked for $\frac{1}{2}$ hour at 150°C.

Acrylic Resin and Formulation. The acrylic resin was obtained commercially (Rohm and Haas QR-496) as a 70% solution in 2-butoxyethanol. The solvent was removed under vacuum, the final conditions being one hour at 120°C and 2 mm Hg. Typically 2000 grams of the resin as supplied yielded 1452 grams of 100% solids resin (72.1% non-volatile). Titration showed the product had an acid value of 36.8. A typical solids determination of the original solution gave 67.5% non-volatile, indicating some residual monomer in the resin.

Coating Formulation (70% non-volatile in 2-Butoxyethanol)

The coating formulation was based on the manufacturer's recommendations (4). A typical formulation used in the second part of the investigation was:

acrylic resin (stripped as above), grams	105
2-butoxyethanol, grams	45
hexamethylolmelamine resin (American Cyanamid XM-1116), grams	22
dimethylethanolamine, grams	4.4
deionized distilled water (to make 10% non-volatile), ml	1094

The quantity of amine added is 63% of theory based on the titrated acidity. Both the amine and the melamine resin were dissolved in the acrylic solution prior to the addition of the water which was accomplished in the same manner as for the epoxy ester. During preliminary studies the hexamethylolmelamine was omitted and the water correspondingly reduced. The electrodeposited films were baked for $\frac{1}{2}$ hours at 175°C.

Trimellitic Based Alkyd and Formulation. The trimellitic based alkyd was supplied as 100% non-volatile material. The acid number as received was 43.2. Calculation showed (*see* below) that the degree of base neutralization recommended was 85% of theory. The typical charge for resin preparation was described as (5):

	Parts by Weight, grams	Molar Ratios
safflower oil	460	2.2
hydrogenated bisphenol-A	347	6
litharge	0.2	
isophthalic acid	80	2
trimellitic anhydride	139	3
	1026	
water of esterification	−26	
	1000	

The resin is prepared in an inert atmosphere. The hydrogenated bisphenol-A and safflower oil are charged, heated to 400°F, and the litharge added. The kettle is then heated to 465°F and held approximately 1½ hours for good alcoholysis. The isophthalic acid is then added, and the temperature is held at 465°F for an acid number of *ca.* 10. The charge is cooled to 380°F, the trimellitic anhydride is added, and the mixture is heated and held at 400°F for an acid number of 40–44. Any solvents are added at 325°F.

Coating Formulation (70% non-volatile in 2-Butoxyethanol)

This formulation, typical of all the combinations, consists of:

trimellitic alkyd, grams	105.0	triethylamine, grams	8.16
2-butoxyethanol, grams	45.0	deionized water, ml	857

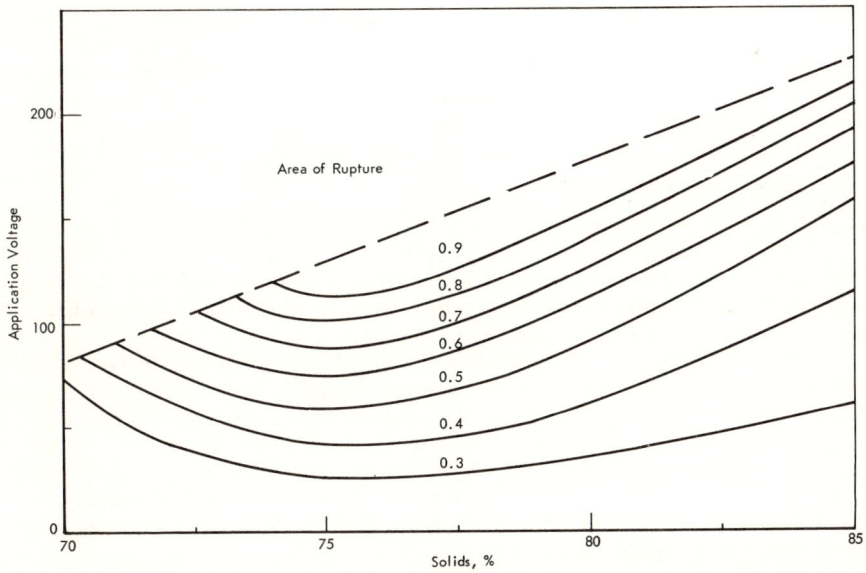

Figure 15. Trimellitic alkyd–2-butoxyethanol: effect of voltage and solids on film thickness (mils)

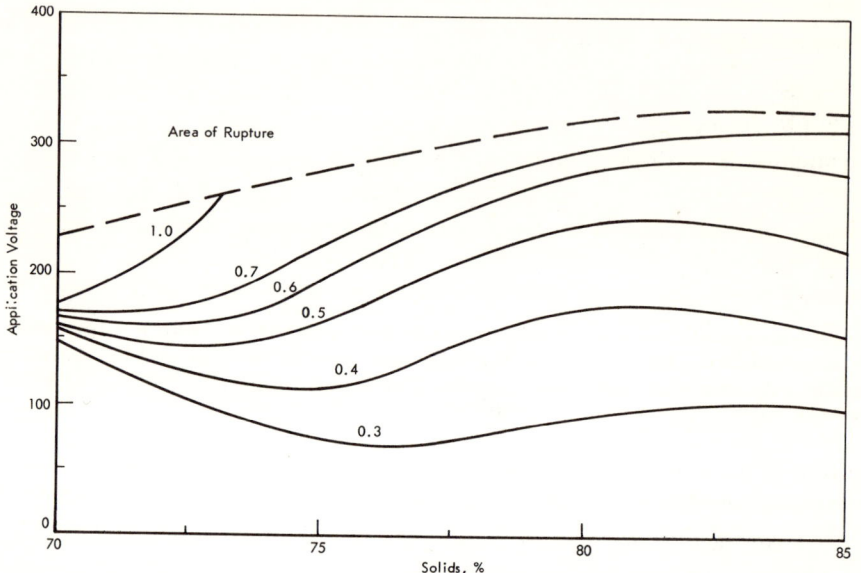

Figure 16. Trimellitic alkyd–2-ethoxyethanol: effect of voltage and solids on film thickness (mils)

The resin was dissolved in the solvent, the amine was added, and the water was added slowly with vigorous stirring to prepare the aqueous solution. The electrodeposited coatings were baked for 20 minutes at 175°C.

Deposition Procedure

All of the coating was carried out on 3 inch × 5 inch, 24-gage zinc-phosphated (Bonderite EP-2), cold rolled steel panels with a deionized rinse. Each panel was immersed in the electrodeposition bath to a constant depth of 10 cm so that a current density of 2.0 ma/cm² could be used for all coating experiments.

The rectifier was a kilovolt model KVI-5000S, modified so that both voltage and amperage could be preset at limiting values. The rectifier was capable of automatic crossover; during the early stages of a run while film resistance was low, it was amperage controlling, and as the resistance built up, voltage eventually became the limiting factor. The amperage limitation permitted simulation of commercial operations and avoided scorching of the coating. Figure 2 shows a typical operation at the rectifier in graphic form.

The voltage and amperage were recorded as a function of time using a Varian G-2000 dual pen recorder. The amperage circuit was equipped with a ball disk integrator permitting a direct readoff of the coulombic consumption. After the panel was coated, the current was shut off, the panel was removed, rinsed thoroughly in deionized water, and the residual water was blown off to avoid spotting. The coating was immediately baked as indicated.

Conclusions

Two solvents used in this evaluation, 2-butoxyethanol and 2-(2-butoxyethoxy)ethanol, were the most versatile. Except for the film appearance of the trimellitic alkyd, satisfactory coating systems could be made with each of the three resins. The 2-butoxyethanol has slightly lower viscosity and lower cost. On the other hand 2-(2-butoxyethoxy)-ethanol gave slightly higher rupture voltages. With the acrylic resin and

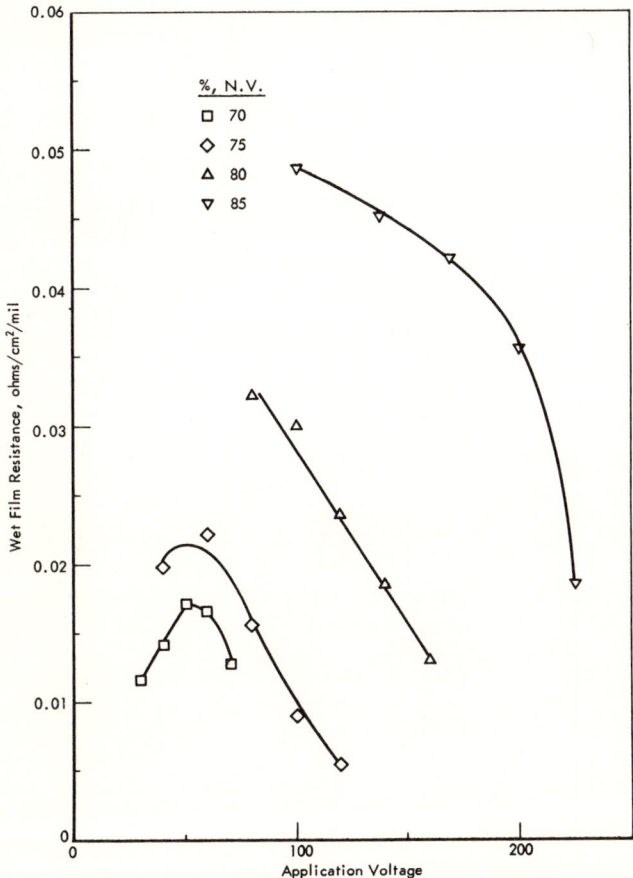

Figure 17. Trimellitic alkyd–2-butoxyethanol: voltage vs. wet film resistance

the alkyd, 2-ethoxyethanol was superior to 2-butoxyethanol. It could not, however, be used with the epoxy ester.

Hexylene glycol was better than 2-butoxyethanol as a solvent for the trimellitic alkyd. It may also be of value with the acrylic resin, but high solution viscosity could be a problem.

Sec-butyl alcohol offered some promise with the trimellitic alkyd and the epoxy ester vehicles, judging from the preliminary phase of the investigation. Further work is needed to establish this fact. An approach involving mixed solvent systems is suggested. With the acrylic resin, the solution properties and deposition characteristics were poor. Isopropyl alcohol offered some promise with the alkyd and the acrylic resin, but further tests are needed to determine its ultimate utility. Diacetone alcohol may be of value with the alkyd and the acrylic. With these resins it

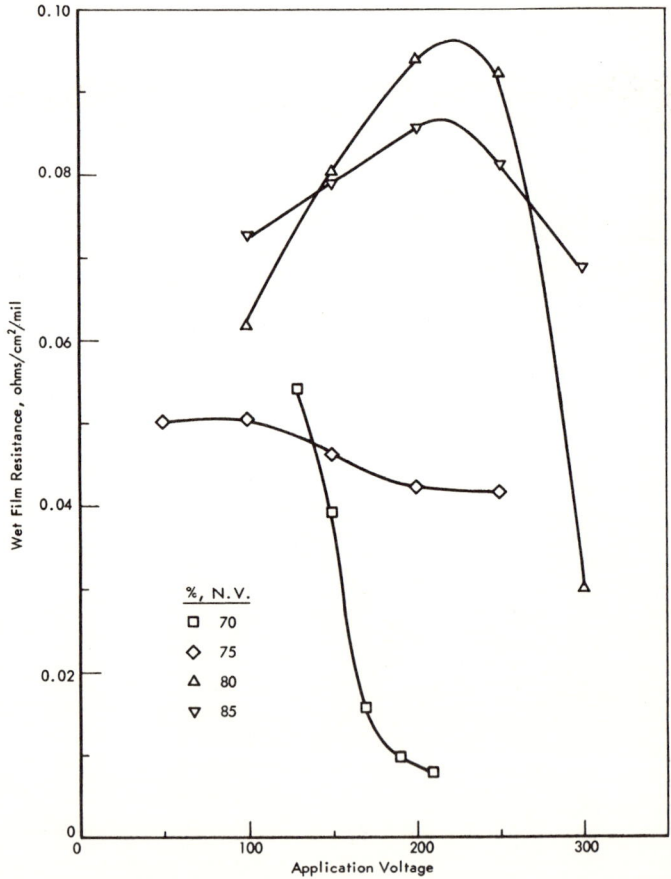

Figure 18. *Trimellitic alkyd–2-ethoxyethanol: voltage vs. wet film resistance*

was one of the best in rupture voltage. It may also contribute something to the cure of the acrylic resin. With methyl isobutyl carbinol coatings could be deposited only with the epoxy ester. Although these coatings were poor in surface appearance, the rupture voltages were such that solvent blends may be worth considering. Methyl isobutyl ketone was ineffective with all three resins.

The results point to several areas worth additional study. In particular, solvent blends should be considered. Since surface appearance was a general problem, more should be known about the residual solvent concentration in electrodeposited films. Although there are procedures to determine the solvent concentration in the deposition bath there have been no reported accurate determinations of the concentration in the electrodeposited film, to the author's knowledge.

Literature Cited

1. Tsou, I. H., Stuecken, W. C., *Amer. Chem. Soc., Div. Org., Coatings Plastics Chem.,* **Preprint 31** (1), 366 (Los Angeles, March-April, 1971).
2. Krylova, I. A., *et al., Lakokrasoch. Mater. Ikh Primen.* **1970** (4), 39.
3. Nunn, C. J., private communication.
4. *Tech. Bull.* **RD-24,** "Experimental Resin QR-496," Rohm and Haas Co.
5. *Bull.* **TMA 102b,** "Experimental Electrocoating Resins from AMOCO TMA," Amoco Chemicals Corp.

RECEIVED March 14, 1972.

11

Polyamide Resin Solubility

E. R. HINDEN and P. D. WHYZMUZIS

General Mills Chemicals, Inc., 2010 E. Hennepin, Minneapolis, Minn. 55413

The solubility of a typical polyamide, flexographic, printing-ink resin is shown in industrially important, monofunctional alcohols, combinations of these alcohols, and some hydrocarbon or hydrocarbon–acetate blended substitutions of these alcohols. The solubility data are plotted on a single chart showing the entire resin concentration solubility range vs. solvent solubility parameter. The conclusion is that methanol has the highest hydrogen bonding force interaction with the resin compared with ethanol, propanol, or 2-propanol and that ethanol provides more combined dipole moment and hydrogen bonding force interaction with the resin than either 2-propanol or propanol.

This work uses the solubility parameter concept to describe the solubility of a typical polyamide ink resin used in flexographic printing. The entire resin concentration solubility is described by coordinates of resin solubility concentration *vs.* solvent solubility parameter. An interpretation of the interaction between the dispersion force, dipole moment, and the hydrogen bonding force of the polyamide resin and industrially important alcohols is made from the data.

Solubility Parameter

Hildebrand (1) showed that a numerical value for the solubility parameter (δ) can be calculated from the equation:

$$\delta = \frac{\Delta E^V}{V^L}$$

where ΔE = energy of vaporization, and V^L = molal volume. Hildebrand also showed that the energy of vaporization, in turn, could be calculated using the equation:

$$\Delta E^{\mathrm{v}} = \Delta H^{\mathrm{v}} - RT$$

where ΔH = heat of vaporization, R = the gas constant, and T = degrees Kelvin.

The solubility parameter or cohesive force of an individual solvent is believed to result from its inner molecular forces of attraction. Individual molecular forces characterize and dominate certain molecular regions of the structure. For instance dispersion (or London) forces result from the association between the electron systems of two adjacent molecules and the arrangement of the electrons. These forces are not affected by temperature, they operate within a short distance, they are accumulative, and they are general. They reside in all molecules and represent the total attractive force known in saturated aliphatic hydrocarbons.

In another example, polar structure gives rise to a dipole moment force. This force is characterized as an electrostatic force, and it can induce weaker dipoles in other molecules. Still another group of forces exists: the Lewis forces which include the exceptional forces developed by hydrogen bonding. The polyamide resins studied here are known for their strong polar and hydrogen bonding forces. Accordingly, they are soluble in solvents with these same strong forces.

Polyamide Resin Solubility

Linear polyamide resins are prepared by the reaction of dimer fatty acid and diamine compounds. The resulting resin has frequent inner molecular ring structures and a repeating amino amide structure.

To determine the solubility parameter for this type of resin (Table I), the total solubility parameter values for several solvents are required. The solubility parameter values for various solvents are easily found in the literature. There is little controversy over the acceptability of the values listed in separate literature sources. The solubility parameter values for solvents of interest here are identical in three popular literature sources (3, 4, 5). A fourth (2) source lists values which have a maximum difference of 3% for some of the alcohols. The list from the popular three applies in this work (Table II).

Solvents of analytical grade or better were used. Solutions at 45 wt % or less solids were shaken on a paint shaker. Systems containing over 45 wt % solids required heating to complete the solution. Tables III, IV, and V contain the solubility data.

The tables describe the resin solubility in two ways. The resin was considered soluble if the resulting solution was clear and fluid. It was designated insoluble if the resulting solution was (a) clear and gel, (b)

Table I. Polyamide Resin Specifications

Color, Gardner, maximum [a]	7
Softening point, °C, ASTM E-28, modified	105-115
Viscosity, poise, 160°C (2)	2.0-5.0

Typical Properties

% Ash	0.01
Color, Gardner	5+
Flash point, °C, ASTMD-92	310
Softening point, °C	112
Specific gravity, 25°/25°C	0.975
Pounds per gallon, 25°C	8.1
Viscosity, poise, 160°C	3.1
Viscosity, 25°C [a]	
Gardner-Holdt	A1-A
Zahn G2, sec.	25
Zahn G3, sec.	11

[a] 40% solids in 99% isopropyl alcohol/super naphtholite (1:1 by weight).

Table II. Solubility Parameters for Specific Solvents

Solvent	Solubility Parameter
n-Heptane	7.4
Cyclohexane	8.2
Isopropyl acetate	8.4
n-Propyl acetate	8.8
Toluene	8.9
Ethyl acetate	9.1
Methyl acetate	9.1
Diethylene glycol	9.1
Diethylene Cellosolve	9.6
Nitrobenzene	10.0
Acetone	10.0
1-Octanol	10.3
2-Nitropropane	10.7
Methyl Cellosolve	10.9
n-Pentyl alcohol	10.9
Nitroethane	11.1
Isopropyl alcohol	11.5
Acetonitrile	11.8
n-Propyl alcohol	11.9
Nitromethane	12.7
Ethanol	12.7
Propylene carbonate	13.3
Methanol	14.5
Glycerol	16.5

cloudy, or (c) separated. The solubility data of the polyamide resin in solvents for flexographic applications include the low molecular weight, monofunctional alcohols and these same alcohols diluted with hydrocarbon or hydrocarbon–acetate. The important aspects of the resin solubility in these three types of solvent are shown in Figure 1.

The solid curve outlines the resin concentration solubility in single and mixed alcohols. Alcohols with delta values on the lower side of the solid curve produce gelled or separated systems; alcohols with high delta values will not dissolve the resin. Although dissolution is completed with methanol at a resin concentration between 60 and 70%, the alcohol is dissolved in the resin in this case.

Table III. Alcohol Solubility[a]

Alcohol	Fractional Volume	δ	Insoluble Solute, wt %	Soluble Solute, wt %
1-Octanol		10.3	45	0–40
1-Pentanol		10.9	60	0–55
Isopropyl alcohol		10.9	60	0–55
Ethanol		02.7	70	0–65
n-C_3H_7OH/C_2H_5OH	.50/.50	12.3	65	0–60
C_2H_5OH/CH_3OH	.75/.25	13.2	18, 75	19–70
C_2H_5OH/CH_3OH	.50/.50	13.6	35, 75	36–70
C_2H_5OH/CH_3OH	.249/.751	14.1	50, 75	55–70
CH_3OH		14.5	55, 75	60–70

[a] Tables III, IV, and V data are graphed in Figure 1. The data represent the boundary solubility points established with more than 300 alcohol–alcohol and alcohol–hydrocarbon solvent–resin combinations.

The discontinuous curves show the solubility extension at lower solubility parameters using certain solvents mixed with alcohol. For example, the maximum solute concentration with isopropyl alcohol (11.5 δ) is 58%. By substituting increasing amounts of heptane until the mixed solvent contains only 10% alcohol, resin solubility extends to 7.9 δ. The extension into lower delta values corresponds to a reduction in solids content. Solubility at lower than 11.5 δ is not successful above 58% resin concentration.

Ethanol gives a solution having a maximum of 66% solids resin concentration. By partially substituting heptane for ethanol, solubility penetration to 7.9 δ can result. Again, no increase in resin concentration but just dilution results.

Higher concentration solutions and minimum solution concentrations result from methanol and methanol–heptane. Substituting toluene for heptane in the methanol combinations enables higher concentrations of solutions than does methanol alone, but no minimum solution conditions result.

Alcohol–hydrocarbon blends are often modified with ethyl acetate. Adding ethyl acetate to alcohol–heptane blends (with the exception of

Table IV. Solubilities for Methanol Combinations

Methyl Alcohol–Heptane Solubility

Solute, wt %	Insoluble δ	Soluble δ
75	≤12.0	≥13.0≤14.0
70	≤ 9.8	≥10.4≤14.3
60	≤ 9.0	≥ 9.4≤14.3
50		≤14.0
40	9.0	≥ 9.3≤13.2
30	≤ 9.3≥13.0	≥10.0≤13.0
20	≤ 9.8≥13.2	≥10.5≤12.0
10	≤11.4≥13.2	12.3

Methyl Alcohol–Toluene Solubiliy

Solute, wt %	Insoluble δ	Soluble δ
75	12.5	13.4
70	10.7	11.
60		10.0
50		9.4
40		9.2
30	14.1	9.5
20	8.9	9.2
10		9.2
5	≤8.9≥13.5	≥9.1≤12.7

Table V. Solubilities for Heptane Combinations

Ethyl Alcohol–Heptane Solubility

Soluble wt %	Insoluble δ	Soluble δ
67	11.9	
65	10.8	11.2
60	9.6	9.8
50	8.4	
40	7.9	8.4
30		8.4
20		10.2
10		7.9

Isopropyl Alcohol–Heptane Solubility

Solute, wt %	Insoluble δ	Soluble δ
55	9.9	10.6
50	8.9	9.3
40	7.9	8.4
30	7.9	8.4
20	7.7	7.9
10		7.9

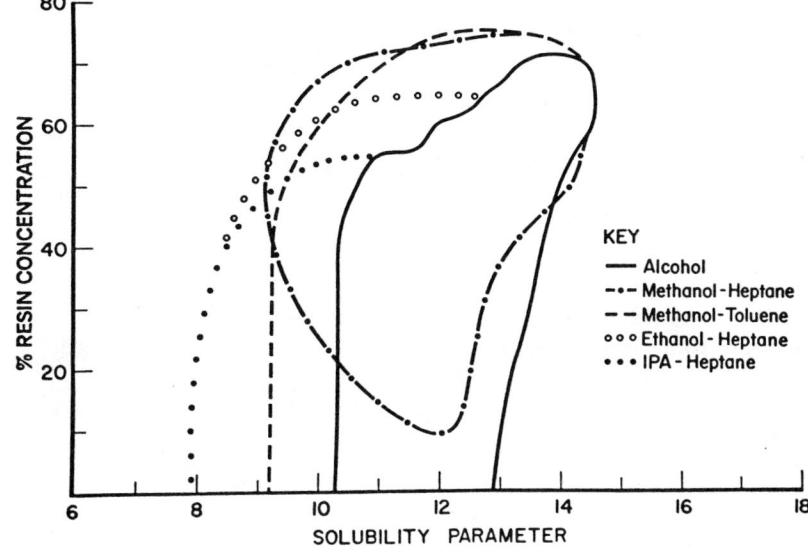

Figure 1. Polyamide resin solubility

methanol blends) decreases solubility in the extended regions. Conversely, methanol–heptane mixtures increase in solvent power to extremes when ethyl acetate is added. To maintain solubility with alcohol–hydrocarbon–acetate blends in the extended region, ethyl acetate should be kept below 40 vol %.

Discussion

The polyamide resin under consideration here has strong hydrogen bonding characteristics and a definite polar structure. Accordingly, monofunctional alcohols which dissolve this resin form strong hydrogen bonds and have definite polar structure. Methanol has the strongest hydrogen bonding interaction with the resin than the other alcohols studied, but it lacks in dipole and dispersion force interactions. It is soluble in the resin and gives this resin enough fluidity to produce non-gelled solutions. When heptane–methanol is used, the introduction of dispersion force increases the maximum solids solution value to 75% and expands the solid minimum to 10%. Combinations of toluene and methanol, which add both dispersion force and dipole moment force, do not increase solution concentration but do remove the minimum concentrations.

Ethanol gives a better combination of dipole and hydrogen bonding forces with the resin than either *n-* or isopropyl alcohol. This trend seems to be amplified where increasing amounts of heptane (dispersion force) replaced alcohol. For instance, where isopropyl alcohol produces gel at just above 58 wt % solids and 11.5 δ, ethanol–heptane gives fluid solutions. Moreover, ethyl alcohol–heptane gives fluid solutions at concentrations as high as 65% at 11.5 δ. None of the alcohols provide the complete balance of forces required for the widest possible range of resin solubility. Ethyl alcohol comes the closest.

Conclusion

The solubility parameter is a useful tool for formulating flexographic inks with polyamide resin. The concept provides insight into the characteristics of inner molecular forces of both polyamide resins and solvents. With the information outlined, the formulator can balance the proper solvent forces by using solvent blends and thus provide latitude in his formulating work. In addition to getting the best solubility, he can incorporate other parameters including cost, drying properties, film performance, and the like to get the best possible solution combinations. A separate solubility graph is required for each resin.

Literature Cited

1. Hildebrand, J., Scott, R., "The Solubility of Nonelectrolytes," Van Nostrand-Reinhold, New York (1950).
2. Hansen, C. M., *J. Paint Technol.* (1967) **39**, 104–117.
3. Crowley, J. D., Teague, Jr., G. S., Lowe, Jr., J. W., *J. Paint Technol.* (1966) **38**.
4. Nelson, R. C., Figurelli, V. F., Walsham, J. G., Edwards, G. D., *J. Paint Technol.* (1970) **42**.
5. Burrell, H., *Off. Dig. Fed. Soc. Paint Technol.* (1955) **27**, 726.

RECEIVED March 14, 1972.

12

Technique for Reformulating Solvent Mixtures in Epoxy Resin Coatings

GEORGE R. SOMERVILLE and JOHN A. LOPEZ
Shell Chemical Co., Woodbury, N. J. 08096

Like most other synthetic resins, epoxy resins have a limited tolerance for certain solvents. A 40% acetone solution can be prepared with an epoxy resin (mol wt 2900), but a 20% solution cannot. Acetone is not a true solvent because a true solvent provides infinite dilution of the resin. The dilution tolerance may be extended by adding alcohols and aromatic hydrocarbons; many such solvent mixtures have become established in epoxy resin based coatings. The study of alternate solvent mixtures has been systematized and simplified using epoxy resin solubility maps and ternary diagrams. The practical application of this reformulating technique with emphasis on the special effects, such as improvements in drying rate and lower viscosity, are shown.

Air pollution restrictions are now controlling the solvent mixtures used in epoxy resin-based coatings requiring reformulation. In conforming to these restrictions, the study of alternate solvent mixtures has been systematized and simplified by epoxy resin solubility maps and ternary diagrams. This paper presents the practical application of this reformulating technique with emphasis on special effects like improvements in drying rate and lower viscosity.

Description of Solubility Maps

Solubility maps usually show the soluble area of a resin in a variety of solvents and are usually based on the physical chemical constants of the solvents. We recognize the various solution parameters such as solubility parameter, internal pressure, dipole moment, fractional polarity, or the various measures of hydrogen bonding, but we have chosen solubility parameter (δ), (the measure of all the intermolecular forces present in the solvent) and fractional polarity (p) (the measure of the relative contribution of the polar forces involved) (1, 2) as coordinates for a two-dimensional solubility map. Table I gives the physical chemical constants of the common solvents that are useful in preparing solubility maps.

A solubility map for two epichlorohydrin (ECH)/bisphenol A (BPA) derived epoxy resins (mol wts 900 and 2900) and a high molecular weight (200,000) thermoplastic epoxy (TE) resin commonly used in coatings is given in Figure 1.

All of the resins have a common region of solubility with the high molecular weight TE resin having a more limited soluble area. While epoxy resins (mol wt 900 and mol wt 2900) have similar solubilities, the higher molecular weight resin is insoluble in acetone, methyl ethyl ketone, and ethyl amyl ketone at 25% solids, and the lower molecular weight resin is soluble in these solvents at 10% solids. The solubility map does not indicate solution viscosity, e.g., epoxy resin (mol wt 2900) solutions are always much higher in viscosity at the same solids.

Solution viscosity is proportional to neat solvent viscosity, but there are other factors which determine solution viscosity. For example, methyl isobutyl ketone and toluene have the same viscosity (0.55 cP) (2), but solution viscosities of a resin in these two solvents or various blends of these solvents have a wide viscosity variation. Solvent composition is also an important factor.

Table I. Physical Chemical Parameters of Solvents

Hydrocarbons	Solubility Parameter	Fractional Polarity
Shell Sol B	7.25	0
Hexane	7.30	0
Shell Sol B-8	7.40	0
Tolu Sol 5	7.20	0
Tolu Sol 10	7.70	0
Shell Sol 260	7.20	0
Toluene	8.90	0.001
Special VM&P naphtha	7.40	0
Super VM&P naphtha 66	7.45	0
Ethylbenzene	8.80	0.001
Xylene	8.85	0.001
TS-28	8.50	0.001
Cyclo Sol 53	8.75	0.001
Shell Sol 70	7.15	0
Shell Sol 360	7.45	0
Shell Sol 340	7.50	0
Cyclo Sol 63	8.70	0.001
Shell Sol 140	7.40	0
Cyclohexane	8.20	0
Styrene	8.66	0

Table I. Continued

	Solubility Parameter	Fractional Polarity
Ketones		
Acetone	10.0	0.695
Methyl ethyl ketone	9.3	0.510
Methyl isobutyl ketone	8.4	0.315
Methyl isoamyl ketone	8.3	0.255
Cyclohexanone	9.9	0.380
Ethyl amyl ketone	8.2	0.223
Pentoxone solvent	8.5	0.190
Diisobutyl ketone	7.8	0.123
Diacetone alcohol	9.2	0.312
Isophorone	9.1	0.190
Esters		
Ethyl acetate	9.1	0.167
Isopropyl acetate	8.6	0.100
n-Propyl acetate	8.75	0.129
sec-Butyl acetate	8.2	0.082
Isobutyl acetate	8.3	0.097
n-Butyl acetate	8.55	0.120
Amyl acetate	8.45	0.067
Methyl amyl acetate	8.2	0.050
Isobutyl isobutyrate	7.7	0.042
Ethylene glycol monomethyl ether acetate	9.2	0.095
Ethylene glycol monoethyl ether acetate	8.7	0.073
Ethylene glycol monobutyl ether acetate	8.2	0.060
Ethers and Glycol Ethers		
Tetrahydrofuran	9.9	0.075
Dioxane	10.0	0.006
Menthyl Oxitol glycol ether	10.8	0.126
Oxitol glycol ether	9.9	0.086
Butyl Oxitol glycol ether	8.9	0.048
Methyl Dioxitol glycol ether	10.2	0.058
Dioxitol glycol ether	9.6	0.043
Butyl Dioxitol glycol ether	8.9	0.028
Diethyl ether	7.4	0.033
Alcohols		
Methyl alcohol	14.5	0.388
Neosol proprietary solvent	13.2	0.296
Isopropyl alcohol	11.5	0.178
n-Propyl alcohol	11.9	0.152
sec-Butyl alcohol	10.8	0.123
Isobutyl alcohol	10.7	0.111
n-Butyl alcohol	11.4	0.096
Methyl isobutyl carbinol	10.0	0.066
2-Ethylhexanol	9.5	0.042
Cyclohexanol	11.4	0.075

Table I. Continued

	Solubility Parameter	Fractional Polarity
Chlorinated Solvents		
Dichloromethane	9.7	0.12
Chloroform	9.3	0.017
1,2-Dichloroethane	9.8	0.043
1,1,2,2-Tetrachloroethane	10.4	0.092
Carbon tetrachloride	8.6	0
1,1,1-Trichloroethane	8.5	0.069
Trichlorethylene	9.2	0.005
Chlorobenzene	9.5	0.058
Miscellaneous		
Diethyl sulfide	8.8	0.065
Carbon disulfide	10.0	0
Dimethyl sulfoxide	13.4	0.813
Dimethyllformamide	12.1	0.772
Nitromethane	12.6	0.780
Nitroethane	11.1	0.710
2-Nitropropane	9.9	0.720
Nitrobenzene	10.0	0.625

The solubility map does not indicate evaporation rate. Evaporation rate depends on many factors including solvent vapor pressure, hydrogen bonding, solvent–resin forces, solvent molecule shape, and polar forces among the constituents of the solution; but solvent blends at the border of immiscibility evaporate more rapidly than blends of true solvents which fall at the center of the soluble area of a solubility map. This is the result of all the factors that influence evaporation.

Ternary Solubility Diagrams

Formulation of solvent combinations for epoxy resin coatings is based on blends of solvents as composites with (δ) and (p) within the solubility area shown in Figure 1. Evaporation rate, flash point, air pollution restrictions (if appropriate), and cost are among the other considerations in solvent selection.

It is useful to formulate solvents using solubility diagrams based on the volume composition of the blends in place of calculating the physical chemical constants for the composites. Ternary diagrams can be readily plotted on three-component solvent blends. Blends containing more than three materials can be plotted on these same diagrams providing the ratio of several of the materials is held constant and the result is only three variables.

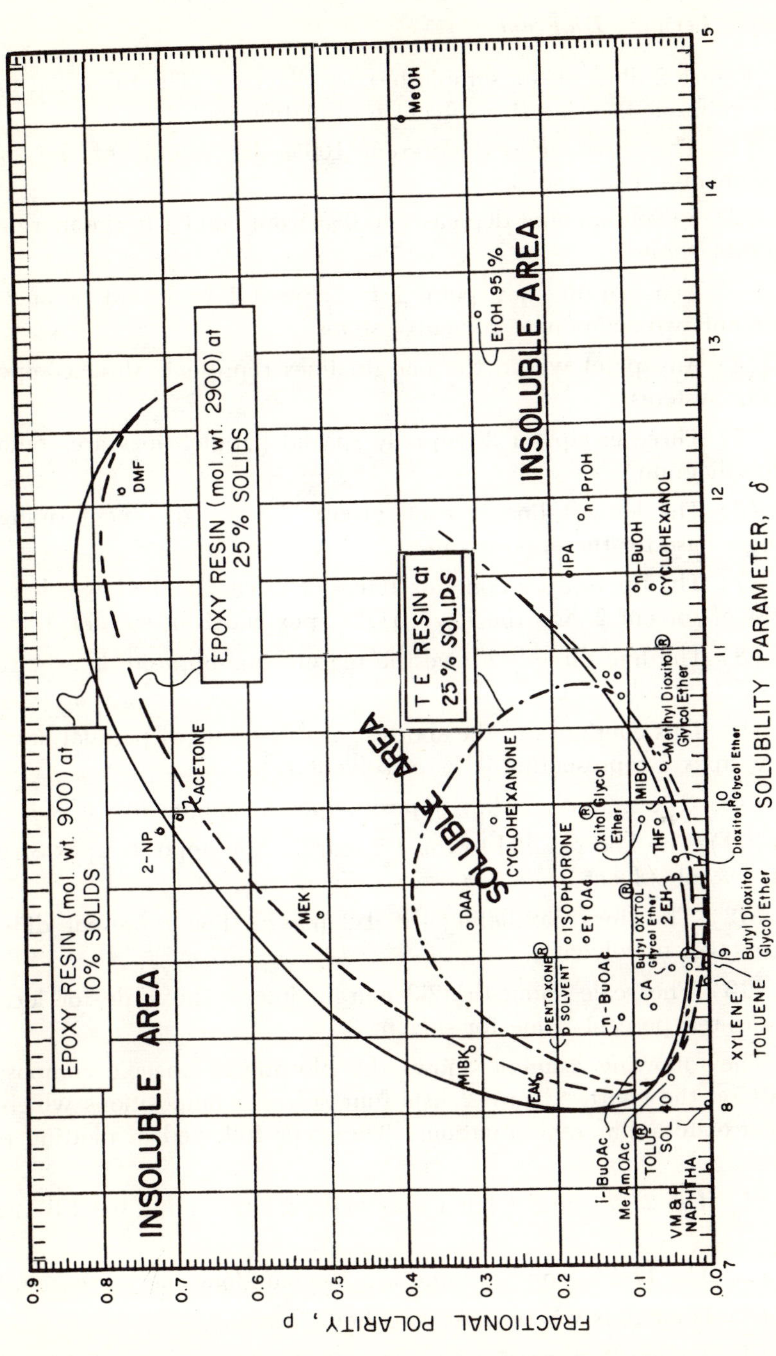

Figure 1. Solubility map for Epon resins at 25°C

Sample Ternary Diagram

Figure 2 illustrates a sample ternary diagram. The following factors are significant in using this solvent formulating tool:

(1) The three apexes represent 100% by volume of the solvent components—1, 2, and 3.

(2) Solvent amount decreases to 0% from point 1 to the margin line opposite point 1.

(3) The margin lines, point 1 to 2, point 1 to 3, and point 2 to 3 represent two-component solvent systems.

(4) Any point within the margin lines represents three-component solvent systems.

(5) Three groups of 20 equally spaced parallel lines are contained in the diagram.

(6) The longest line in each group is a margin line forming the triangle base for the opposite apex.

(7) The upper apex contains 100% of solvent 1, the lower left apex 100% of solvent 2, and the lower right apex 100% of solvent 3.

(8) The line directly above the margin line opposite apex 1 represents 5% of solvent 1.

(9) This continues to the shortest of this group of parallel lines just below apex 1 representing 95% of solvent 1.

(10) The 1-2 margin line opposite apex 3 contains 0% of solvent 3.

(11) The line parallel to the 1-2 margin line opposite apex 3 represents 0% of solvent 3.

(12) The lines paralleling the 1-2 margin line represent different percentages of solvent 3.

(13) The scale along the 2-3 margin line is the scale for locating any desired percent value for solvent 3.

The following example shows the plotting of specific composition points on the chart. Figure 2 lists four solvent compositions which are typical of four different situations. The steps followed in plotting point A are:

(1) The 25% solvent 1 line was located intersecting the 1-3 margin line.

(2) The 17½% solvent 2 line was located along the 1-2 margin line.

(3) Point A is where these two lines intersect.

(4) To check, determine the amount of solvent 3 specified by point A by following the line through point A parallel to margin line 1-2.

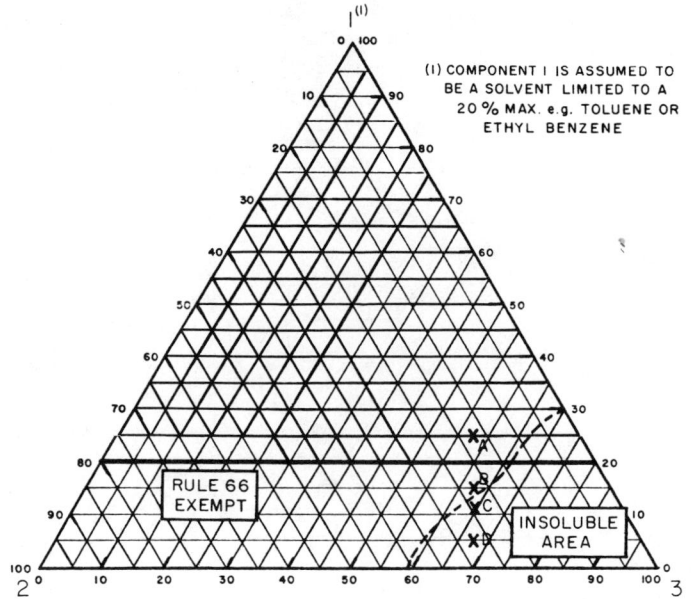

Figure 2. Sample ternary diagram.

(5) The amount of solvent 3 is where this line intersects margin line 2-3.

(6) The percentages of solvent 1, 2, and 3 at point A, or any other point, must total 100.

Ternary Solubility Diagram of Epoxy Resins
(mol wt 900 and mol wt 2900)

The solubility map for epoxy resins (Figure 1) shows that a soluble region should result from blends of Oxitol glycol ether, toluene, and isopropyl alcohol. Oxitol glycol ether is the only true epoxy resin solvent in this blend. Combining these solvents results in a sizable area of solubility as shown in Figure 3.

Epoxy resin (mol wt 2900) is less soluble than epoxy resin (mol wt 900), but only 20% isopropyl alcohol or Oxitol glycol ether is needed

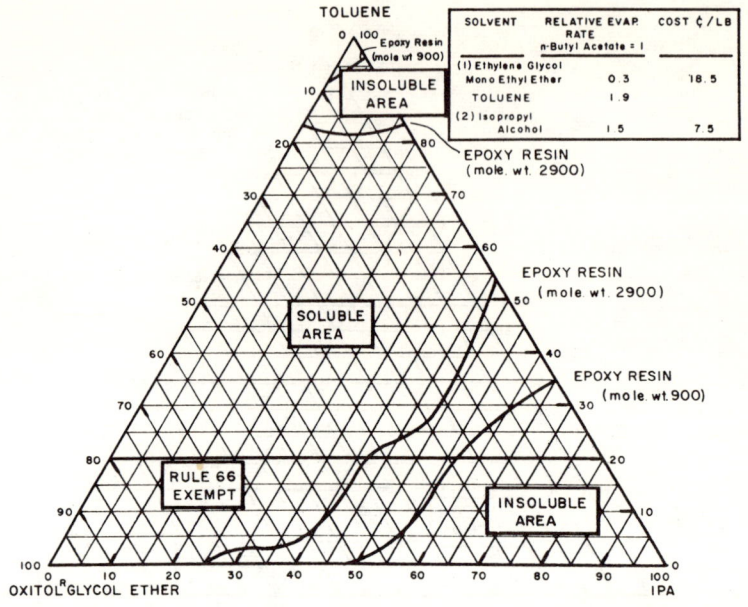

Figure 3. Oxitol glycol ether/toluene/IPA systems. Solubility of epoxy resin, mol wt 900. Resin solids are 25 vol %.

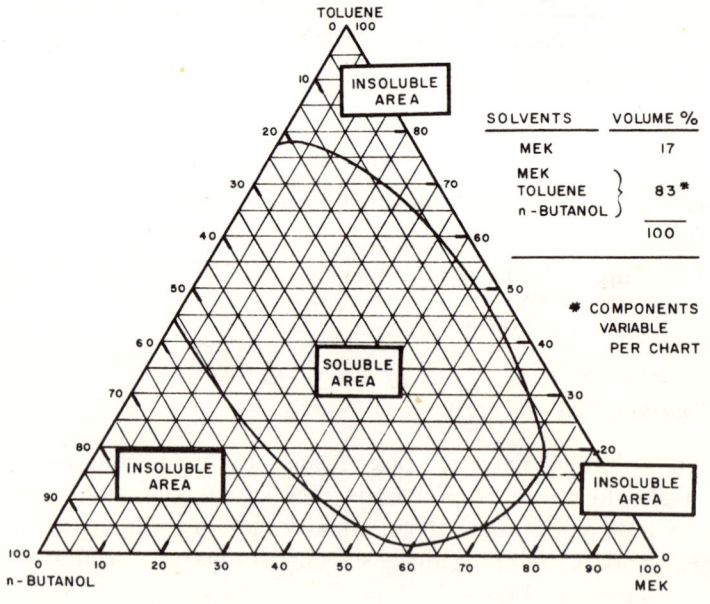

Figure 4. TE resin solubility. MEK/toluene/n-butyl alcohol at 10 vol % solids.

with toluene to dissolve the resin. With the lower molecular weight epoxy resin (mol wt 900), 90–95% toluene in the solvent blend gives a 25% solids solution.

TE Resin Solubility

Figure 4 illustrates ideally the use of the solubility map for epoxy resins (Figure 1). Toluene, methyl ethyl ketone, and n-butyl alcohol are not solvents for the TE resin at 10% solids, but a considerable portion of the soluble area for this resin is within the boundary limits of these three non-solvents. A large soluble area is formed by combinations of these three non-solvents as seen in Figure 4.

Special Effects Accomplished by Formulating with Ternary Solubility Diagrams

The experimental, flexibilized thermoplastic epoxy (FTE) resin ternary solubility diagrams given in Figures 5 and 6 demonstrate the usefulness of these formulating tools in achieving special effects such as lowering viscosity or improving drying rate.

The FTE resin as formulated contains 40% solvent (28/15:acetone/toluene) by volume. Figure 5 indicates the solubility and viscosity profile of this resin when reduced to 40% non-volatile (application solids) with methanol, hexane (Shell Sol hydrocarbon solvent B), Super VM & P naphtha, or blends of these solvents.

It can be seen that:

(1) Immiscible systems result when either methanol or hexane is the sole addition solvent.

(2) Super VM & P naphtha addition results in a soluble system, and

(3) Maximum solution viscosity is obtained using Super VM & P naphtha whereas minimum solution viscosity results from methanol, hexane blends, i.e., blends of the two non-solvents reduce viscosity to a greater degree than the use of a single true solvent. It was also observed, in studying the drying rate of the resin, that slow dry is related to the solvent retentive nature of the polymer. Non-solvents are not retained by the polymer.

Solubility work shown in Figure 6 shows that solution can be achieved by two non-solvents with the aid of a true solvent, e.g., 55/40/5 vol %:hexane/methanol/toluene or acetone. Toluene or acetone is the only true solvent in the blend. Paints were made with the following two solvent combinations to demonstrate the value of a nearly immiscible solvent blend:

	% Volume	
	#1	#2
Hexane	55	60
Methanol	40	10
Acetone	5	30
	100	100
Evaporation rate, sec.	94	60

The ternary solubility diagram (Figure 6) shows that the most significant difference between the two solvent blends is the solvent composition route during evaporation, *i.e.*, solvent blend 1 becomes immiscible almost immediately while blend 2 must lose considerable solvent before becoming immiscible. The drying rate of paints containing these two blends confirmed that blend 1 dries much faster than blend 2 even though the evaporation rate of blend 2 is much faster (60 *vs.* 94 sec respectively).

Summary

Following the technique of solvent selection by solubility maps and ternary solubility diagrams, the coatings formulator can adapt solvent blends for epoxy resins to obtain lower viscosities and improved drying rates. It is obvious that lowering the solvent cost, conforming to air pollu-

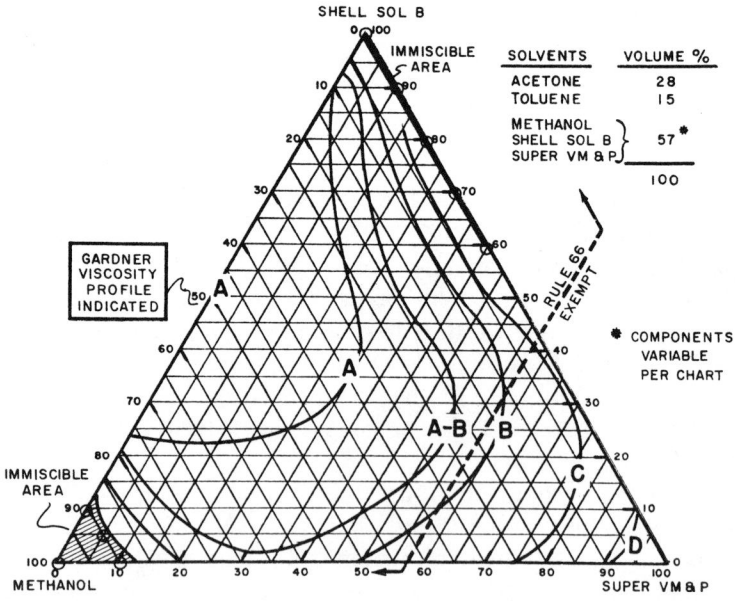

Figure 5. FTE resin. Solubility and viscosity profile (40 vol %, non-volatile).

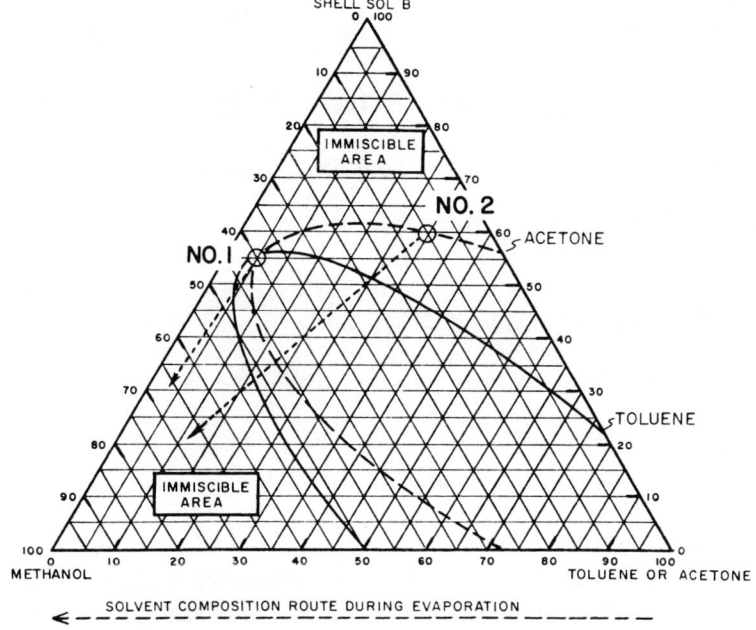

Figure 6. FTE resin solubility map. 40 vol % solids; acetone or toluene with Shell Sol B and methanol.

tion restrictions, and raising the flash point can be accomplished by the same technique.

Literature Cited

1. Gardon, J. L., *J. Paint Technol.* (1966) **38**, 43.
2. Nelson, R. C., Figurelli, V. F., Walsham, J. G., Edwards, G. D., *J. Paint Technol.* (1970) **42**, 644.

RECEIVED March 14, 1972.

13

Solubility Characteristics of Today's Vinyl Chloride Homopolymers, Copolymers, and Terpolymers

RUSSELL A. PARK

Firestone Plastics Co., Box 699, Pottstown, Pa. 19464

The chemical compositions of commercially available vinyl chloride based polymers are discussed with respect to polymer solubility, coating application techniques, and physical properties of the applied coating. Parameters considered include effect of molecular weight on homopolymer solubility; effect of molecular weight on vinyl chloride–vinyl acetate copolymer solubility; unique copolymer compositions such as VC-TFCE; and terpolymers containing reactive carboxyl or hydroxyl groups. Polymer dissolution techniques and their effect on fluid properties are recognized. Considerations relevant to viscosity measurements of non-Newtonian fluids are cited using viscometers operating through a broad range of shearing rates. True or active solvents studied include ketones, chlorinated hydrocarbons, acetates, nitroparaffins, toluene, and xylene.

Homopolymers, copolymers, and terpolymers of vinyl chloride have been used in the protective and decorative coating fields for nearly three decades. Their resistance to acids, alkalies, weathering, salt water, oil, greases, and food products, coupled with their flexibility, toughness, clarity, ease of pigmentation, flame resistance, and low odor and taste characteristics have contributed to their position in the market place.

Proper solvent selection ensures that the vinyl-based polymers can be applied *via* conventional fluid techniques of spread and roll coating, dipping, spraying, brushing, etc. When used alone they behave as conventional lacquers. When used in conjunction with other polymer systems the reactive vinyl polymers permit crosslinking.

Vinyl Chloride

More pounds of homopolymers are produced than any other vinyl resin. Figure 1 shows the typical reaction of monomeric vinyl chloride to form poly(vinyl chloride). Polymers of this nature represent the ultimate in chemical and physical properties.

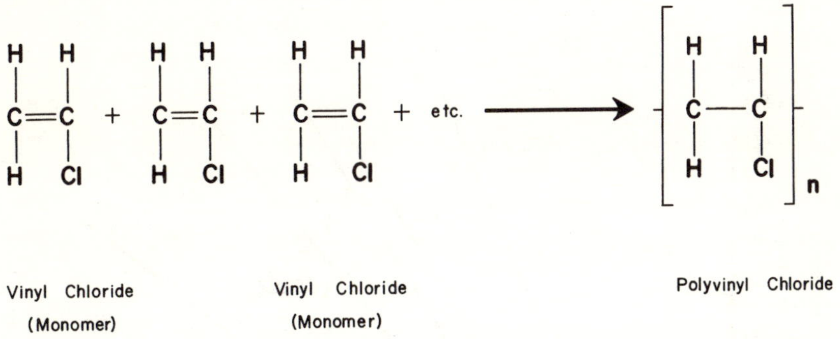

Figure 1. A typical reaction of monomeric vinyl chloride to form poly(vinyl chloride)

Physical properties are directly proportional to molecular weight; however solubility is inversely proportional to molecular weight. The straight PVC [poly(vinyl chloride)] resins have two distinct disadvantages when they are compared with conventional solution resins:

(1) They are not capable of metal adhesion.
(2) They have a much lower order of solubility when compared with the copolymer solution resins. This limits their application because of selection of solvents (for effect upon evaporation or drying rate, and in many cases, cost).

Extremely strong solvents such as THF (tetrahydrofuran) or cyclohexanone are frequently suggested for use with homopolymer resins. Figure 2 shows the typical solubility of these resins in THF.

Figure 3 shows two of these resins in a conventional vinyl solvent, *i.e.*, MEK (methyl ethyl ketone); note their significantly reduced solubility. Because of the limitations cited previously, only a very small segment of the homopolymers of vinyl chloride produced today find their way into the solvent coatings industry. Two of their main applications are in cast films (for packing) and in topcoats for calendered and plastisol (cast) sheet goods.

The most widely produced vinyl chloride copolymers are the resins containing vinyl acetate. Figure 4 shows the typical reaction of vinyl chloride and vinyl acetate to form a copolymer. Vinyl acetate levels are available up to about 15 wt % vinyl acetate on a commercial basis. These polymers give the best chemical and physical properties of the conventional solution resins, and they have much greater solubility than the straight PVC resins. The lower molecular weight series have better aromatic tolerance. The vinyl acetate copolymers do not have metal adhesion although they are compatible with other vinyl resins that contain reactive groups for this purpose. Vinyl-to-vinyl adhesion and vinyl-to-porous substrate adhesion are usually considered excellent.

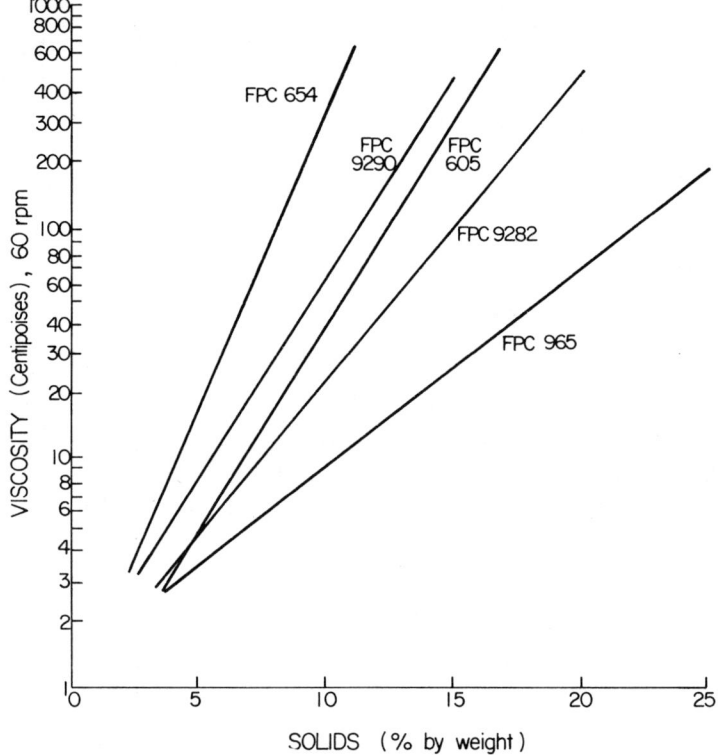

Figure 2. Solubility characteristics of PVC resins in tetrahydrofuran (Brookfield LVF viscometer)

Figure 3 shows the solubility characteristics of the vinyl acetate copolymers in MEK; note the improved solubility over the homopolymers of vinyl chloride. Because these copolymers are still significantly less soluble than the metal adhesion terpolymers, one manufacturer has produced a copolymer not based on vinyl acetate (*see* Figure 3, high solubility copolymers). This copolymer is described in more detail in Figure 13. Copolymers are also available based on vinyl chloride–trifluorochloroethylene and vinyl chloride–vinylidene chloride.

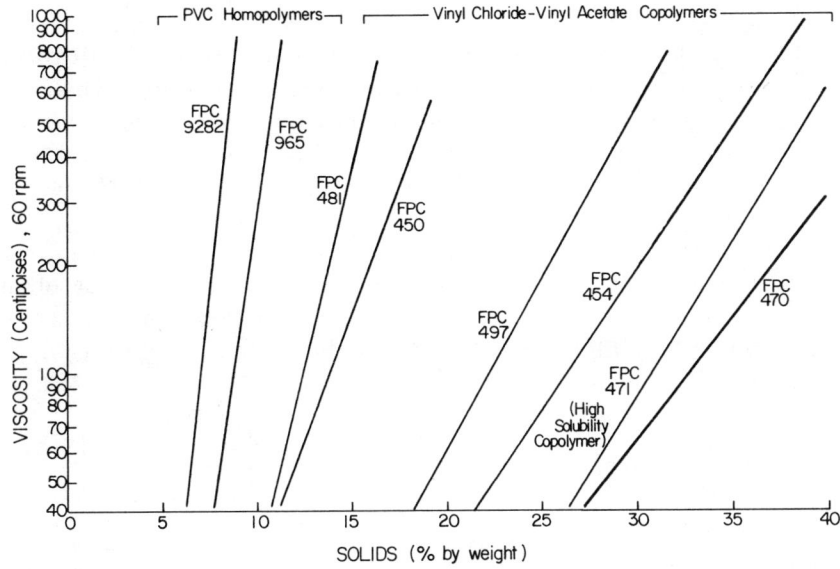

Figure 3. Solubility characteristics of typical vinyl polymers in MEK (Brookfield LVF viscometer)

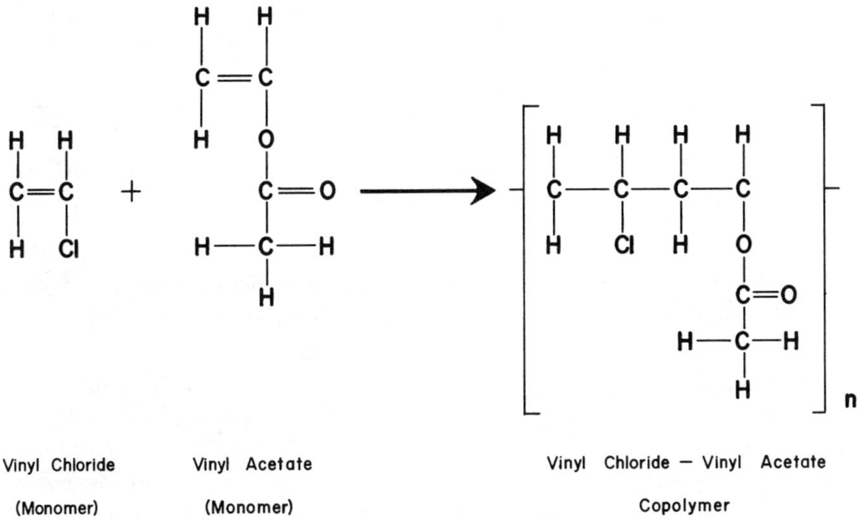

Figure 4. A typical reaction of vinyl chloride and vinyl acetate to form a copolymer

Terpolymers of vinyl chloride containing carboxyl groups or hydroxyl groups are available; the carboxyl groups are added to obtain adhesion to metals. However, we should not expect to obtain the best chemical and physical properties from a resin that is inherently reactive (for metal adhesion). These resins are frequently used in combination with other resins to impart adhesion to systems containing these resins. The converse can also apply, *i.e.*, the resins cited earlier are used to improve the chemical and physical properties of the metal adhesion (reactive) type resins. Metal adhesion type resins are available that are 100% soluble in toluene or xylene.

Because these resins are acidic, care should be exercised in formulating ingredients which are basic. The basic ingredients most often encountered are color pigments. Extensive information regarding color pigment reactivity with these resins is available in the literature (1).

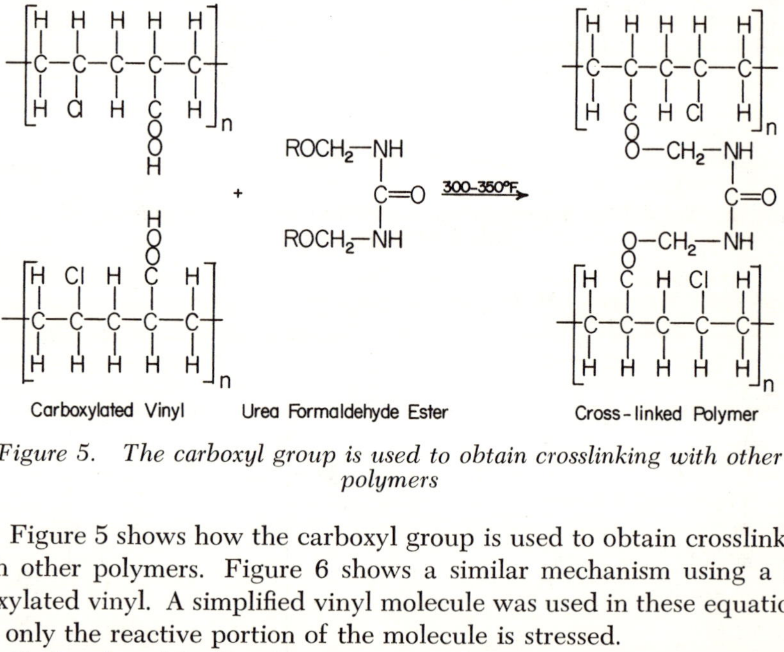

Figure 5. The carboxyl group is used to obtain crosslinking with other polymers

Figure 5 shows how the carboxyl group is used to obtain crosslinking with other polymers. Figure 6 shows a similar mechanism using a hydroxylated vinyl. A simplified vinyl molecule was used in these equations, *i.e.*, only the reactive portion of the molecule is stressed.

Carboxyl and hydroxyl groups are added also to the polymer chain to increase compatibility with other resinous materials such as melamine, alkyds, urethanes, etc. Some cases of incompatibility between polymers may not originate from polymer chemistry but rather from solvent strength. If a vinyl terpolymer in a ketone solvent system is mixed with another polymer system which is dissolved primarily in (for example) an alcohol, the solvent mixture may be too lean for the vinyl which will precipitate out or develop a hazy film.

Crosslinking is a common technique used to develop solvent resistance in the final coating and to increase hardness or heat distortion temperature (necessary if the thermoplastic vinyl coating required sanding during its lifetime). More details regarding crosslinking of vinyl polymers are cited by Reichard (2).

Molecular Weight Classifications. Dilute solution measurements are frequently used to indicate the molecular weight of a vinyl polymer. Expressions like relative viscosity, specific viscosity, inherent viscosity, etc. are used to describe a measurement of 1 or less wt % of resin in a true solvent (3). Preferred solvents for these measurements include cyclohexanone, nitrobenzene, or ethylene dichloride. These measurements have meaning only when polymers of the same composition are measured, i.e., only homopolymers or vinyl chloride/vinyl acetate copolymers with the same level of vinyl acetate, etc. Erroneous results have been encountered when polymers containing post additives or polymerization ingredients which were not soluble in the test solvent were evaluated. In these situations, it may be practical to evaluate the polymer in an instrument which uses a much higher level of polymer such as the torque-rheometer (4). This procedure involves measuring a hot-melt viscosity rather than a dilute solution viscosity. Formulation and evaluation temperature may have to be altered to accommodate the chemical nature of the specific polymer being evaluated.

True Solvents

Before we can discuss solubility characteristics, it is important to describe the technique used to put the polymers into solution and the significance of the viscosity measurement. All solutions were prepared

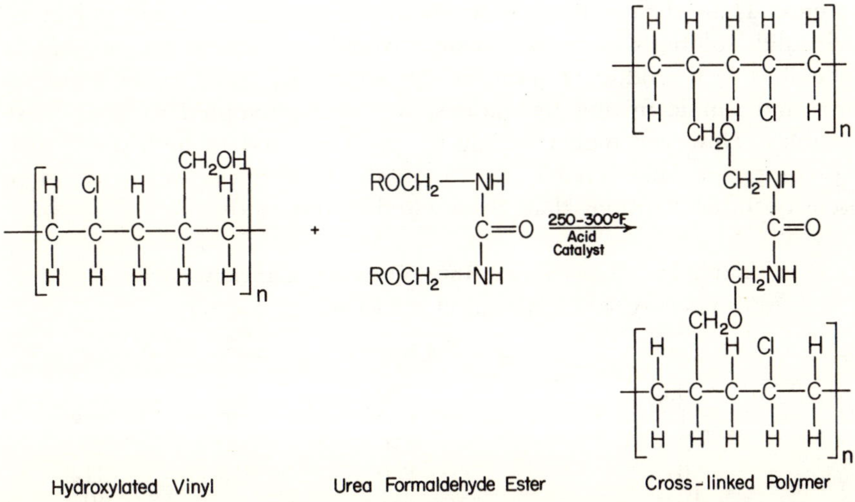

Figure 6. A mechanism for a crosslinked polymer from a hydroxylated vinyl

Table I. Formulation and Physical Properties of Three Vinyl Solvent System Coatings

	Formulation parts by wt		
	Formula I	Formula II	Formula III
FPC 470	30	40	
FPC 450			15
MIAK	80	60	
Pentoxone			85
Dioctyl phthalate	3		
TiO_2 (Ti-Pure R-902)	22.5		
Gold Bond R	15		
Asbestine 325	22.5		
Propylene oxide	0.5		
Bentone 27	1.0		
No. 2 Zahn Cup	200		
No. 4 Zahn Cup	46		
No. 4 Ford Cup	135		
Non-volatile, wt %	53.8	40	15
Non-volatile, vol %	31.4		
Pigment volume concentration, %	43.0		
Pigment/binder	1.9/1		
Lbs/gallon	12.56	8.03	7.94

by adding the weighed resin into a glass jar containing the solvent and by rolling the jar overnight on a ball mill (16 hours). This procedure is the minimal requirement in many cases for obtaining complete solution of the polymer. Park and Klinger (5) have studied the effects of mixing techniques and solvent selection on the viscosity of solution vinyl polymer systems. These data indicate that when high molecular weight polymers, high solids solutions, or weak solvent mixtures are present, we can expect difficulties in reproducing solution viscosities. By using room temperatures and simple mixing techniques, we have attempted to limit these variables. However, higher mixing temperatures and/or high speed mixing equipment may enable us to obtain higher viscosity and more viscosity-stable solutions than those cited in this study.

Table II. Low Shear Rate Viscosity Characteristics
Viscosity, cP vs. Brookfield (LVF) spindle speed, rpm

rpm	Formula I	Formula II	Formula III
0.6	550	620	
1.5	610	632	3680
3	670	640	3600
6	860	649	3420
12			3050
30			2576

Viscosity measurements used to describe polymer solubility are based on the Brookfield model LVF, at 60 rpm after 1 min spindle revolution. This is a rotational type viscometer which can vary its shear rate by spindle speed changes. Only one-point (shear rate) measurements were made however. This type of reporting puts it on a par with the efflux type viscometers used in the coating industry in that only one shear rate is used.

Rheological terms have been described previously (6). When Newtonian fluids are being studied, a single shear rate is adequate. Studies with high solid dispersions (plastisols and organosols) have shown that with non-Newtonian fluids, the apparent viscosity can vary significantly with shearing rate. Solution vinyl systems can also exhibit this behavior. Table I cites formulation and physical properties of three vinyl solvent system coatings. Formulation I is an airless spray, high build, maintenance coating with a pigment/binder ratio of 2. Formulation II is a high-solids clear lacquer (40 wt % in MIAK) based on a COOH containing

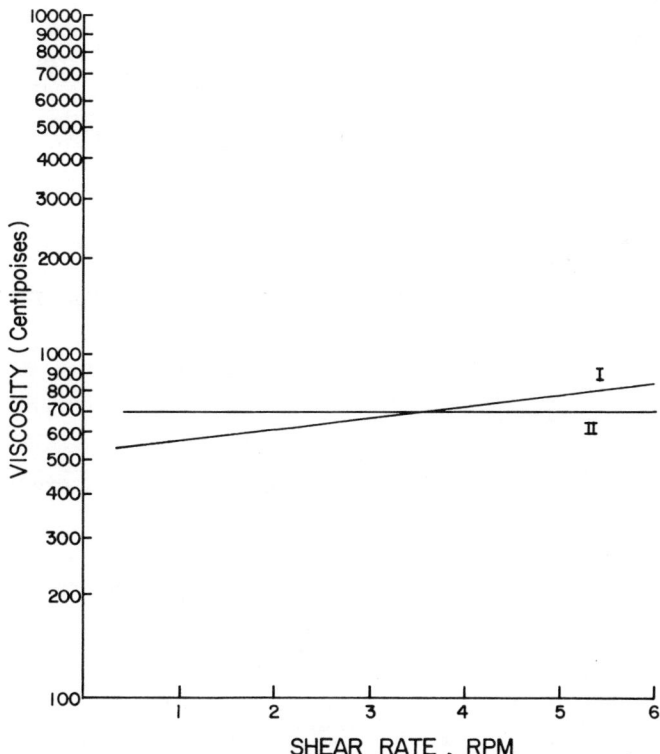

Figure 7. Low shear rate viscosity characteristics of formulations I and II from Table II (Brookfield LVF viscometer)

vinyl terpolymer. Formulation III is 15 wt % solids, clear lacquer based on a high molecular weight vinyl chloride–vinyl acetate copolymer in a ketone solvent (Pentoxone). Efflux type viscosity measures are cited (Ford Cup, Zahn Cup).

Formulas II and III should exhibit Newtonian viscosity characteristics. Formula III with the high pigment/binder ratio might be expected

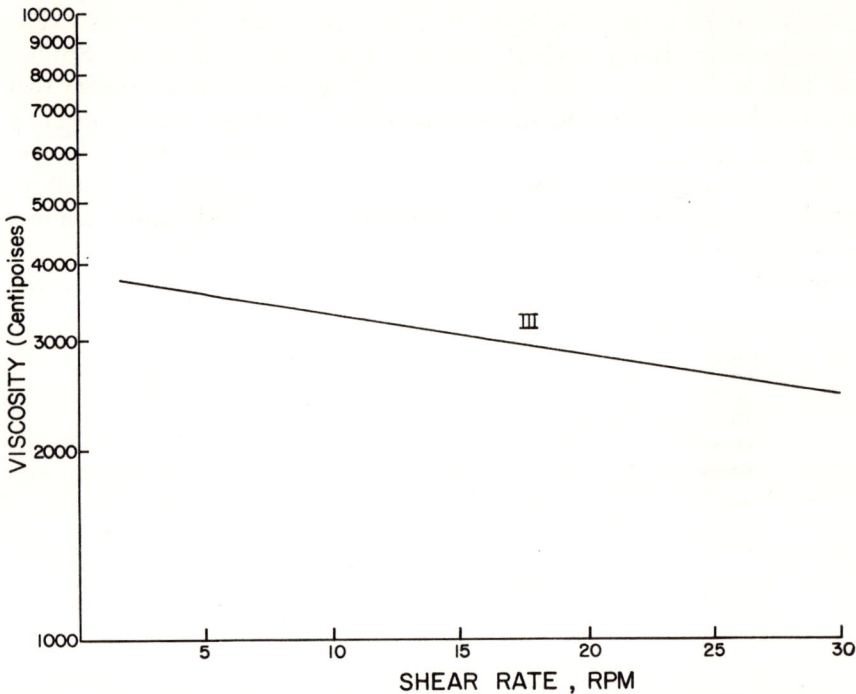

Figure 8. Low shear rate viscosity characteristics of formulation III from Table II (Brookfield LVF viscometer)

Table III. Intermediate Shear Rate Viscosity Characteristics[a]

Viscosity cP vs. Shear Rate (sec^{-1})

Shear Rate, sec^{-1}	Formula I	Formula II	Formula III
		Viscosity (cP)	
19.9	1660	1730	2890
29.9	1300	1440	2430
39.9	1230	1320	2200
62.3	880	1160	1930
93.5	730	1030	1740
125	640	1010	1620

[a] Epprecht viscometer data.

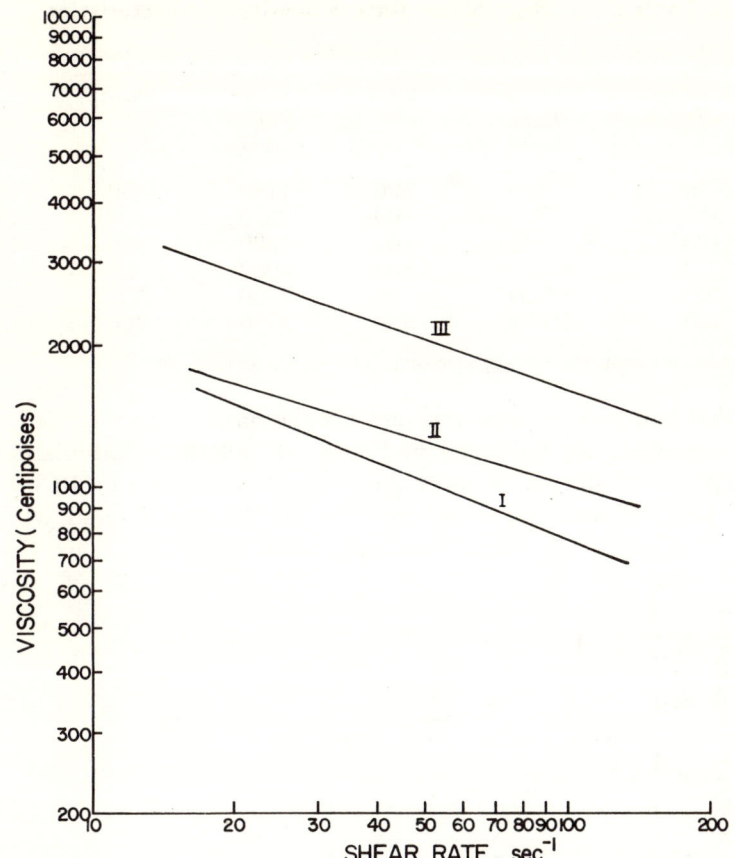

Figure 9. Intermediate shear rate viscosity characteristics of formulations I, II, and III from Table II (Epprecht viscometer)

to show a deviation from Newtonian viscosity. Table II illustrates the viscosity–shear rate relationships of these systems based on the Brookfield LVF viscometer. Data obtained from Formulas I and II are plotted in Figure 7. Formula I under the shear rates studied is dilatant; Formula II is Newtonian. Data from Formula III are plotted in Figure 8; the curve is pseudoplastic.

Table III shows the viscosity characteristics of these systems as obtained from the Epprecht rotational viscometer. With this instrument, shear rate can be obtained in units of reciprocal seconds (sec^{-1}). These data are plotted in Figure 9. Under these shearing rates (19.2 to 125 sec^{-1}), all three formulations are pseudoplastic.

Table IV shows the viscosity characteristics of these systems under high shear rates. The Severs extrusion rheometer, an efflux type instru-

Table IV. High Shear Rate Viscosity Characteristics[a]

Formula I		Formula II		Formula III	
Viscosity, cP	Shear Rate, sec^{-1}	Viscosity, cP	Shear Rate, sec^{-1}	Viscosity, cP	Shear Rate, sec^{-1}
480	2300	900	1180	1100	1090
415	5390	800	2670	800	2970
420	7950	800	4300	600	5800
390	11,500	800	5980	500	8850
420	13,260	700	7750	500	10,990
410	16,300	800	8860	400	15,900

[a] Severs extrusion rheometer, viscosity (cP) vs. shear rate (sec^{-1}).

ment that can vary its shearing rate *via* the gas pressure applied to the fluid, was used. As indicated in Figure 10, all three formulations are pseudoplastic under shear rates that extend from 1090 to 16,300 sec^{-1}. These data illustrate the problems that can be encountered when vis-

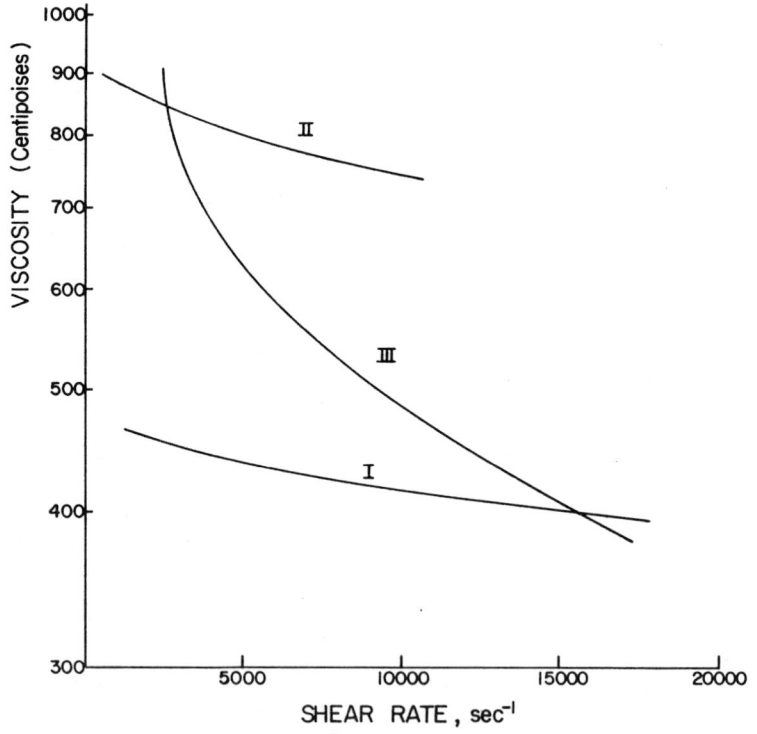

Figure 10. High shear rate viscosity characteristics of formulations I, II, and III from Table II (Serves extrusion rheometer)

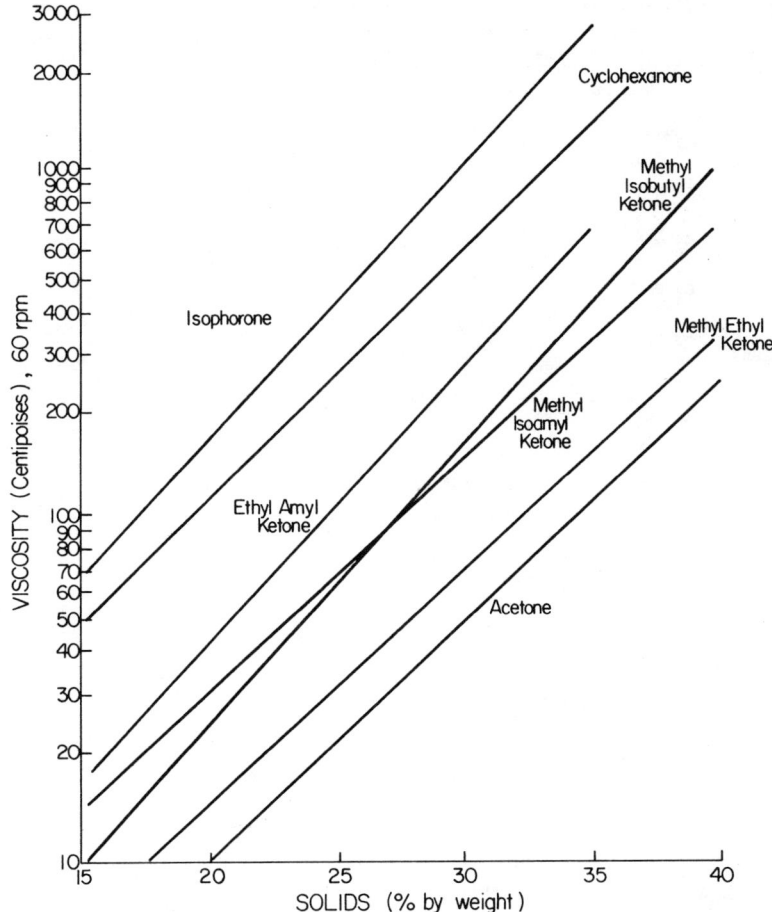

Figure 11. Vinyl terpolymer (FPC 470) in ketone solvents (Brookfield LVF viscometer)

cosity characteristics of non-Newtonian fluids are measured with a viscometer that is unable to vary shear rates or measure viscosity at the shear rate to be encountered in actual application.

Much has been written about ketones, as a class, as excellent solvents for vinyl polymers (7, 8). Aside from solubility of the solvent, we also must consider its volatility characteristics (evaporation rate), flash point, boiling range, etc. With the high solubility type polymers, we have wide latitude with respect to selecting the ketone that will balance these properties against solution solids. This is evident in the next figures.

Figure 11 shows the viscosity characteristics of a vinyl chloride terpolymer which contains carboxyl groups. Excellent solubility, evidenced by high solids, is evident in ketones varying in flash point from acetone to isophorone or cyclohexanone. Figure 12 shows the high solids available with ketones using another carboxyl containing terpolymer.

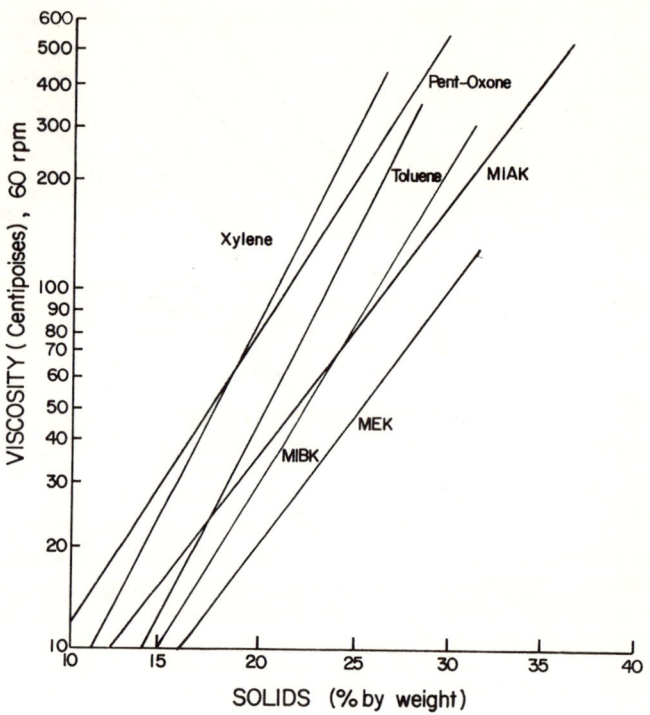

Figure 12. Solubility characteristics of a vinyl terpolymer (Brookfield LVF viscometer)

Figure 13 shows the excellent solubility of a vinyl chloride specialty copolymer in ketones. The fluorinated vinyl chloride copolymer also shows the same excellent solubility in ketones as shown in Figures 14 and 15.

Only the lower molecular weight vinyl chloride–vinyl acetate copolymers approach the aforementioned high solubility vinyl chloride copolymers or terpolymers in solubility. This is shown in the curves in Figure 16, where a low molecular weight vinyl chloride–vinyl acetate copolymer is studied in various ketone solvents. When high molecular weight vinyl chloride–vinyl acetate copolymers are used, solubility drops significantly (Figure 17). This drop in solubility, however, is balanced against improved physical properties (toughness) of the film. The solubility of the homopolymers of vinyl chloride (regardless of molecular weight) is usually poor in most conventional ketone solvents.

The hydroxyl containing terpolymers have solubilities similar to those of the medium-to-low molecular weight vinyl chloride–vinyl acetate copolymers. This resin solubility system is shown in Figure 18. The ketones evaluated were methyl n-butyl ketone and methyl n-propyl ketone.

Many esters such as butyl acetate and ethyl acetate are used as solvents for vinyl polymers. Figure 19 shows the solubility properties of a carboxyl containing terpolymer in various acetate solvents. These esters can also be used as diluents or to alter evaporation rates in mixed solvent systems. If a primary solvent such as a ketone is present, very high solids can be obtained.

Nitroparaffins are effective vinyl solvents. The solubility characteristics of 2-nitropropane are shown in Figures 20 and 21. Toluene and xylene are true solvents for only the most soluble copolymers and terpolymers. These polymers are represented by the carboxyl containing terpolymer (Figure 12), the high solubility copolymers (VC–TFCE copolymer) (Figure 14), and a non-vinyl acetate containing copolymer (Figure 13). Heat had to be applied to the last polymer to obtain complete solution.

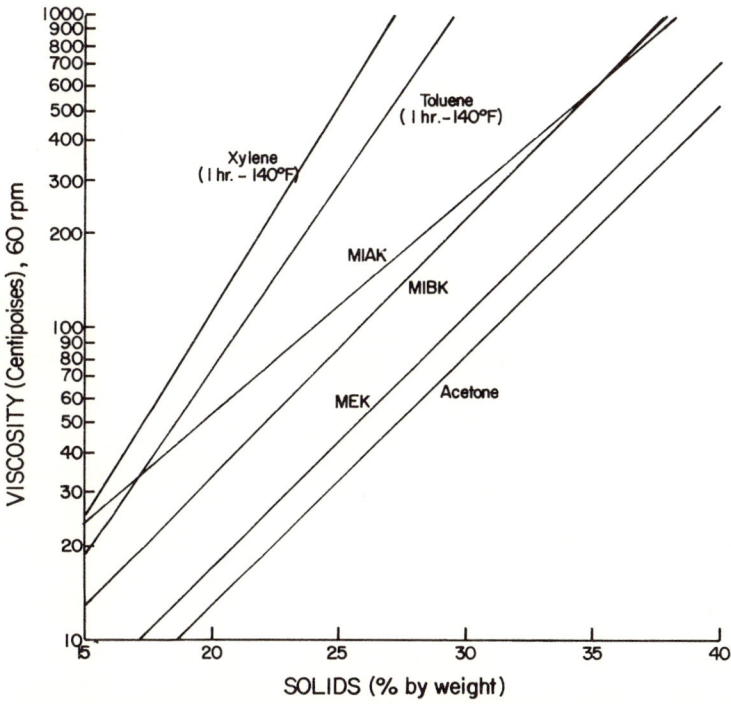

Figure 13. Solubility characteristics of a vinyl chloride specialty copolymer (FPC 471) (Brookfield LVF viscometer)

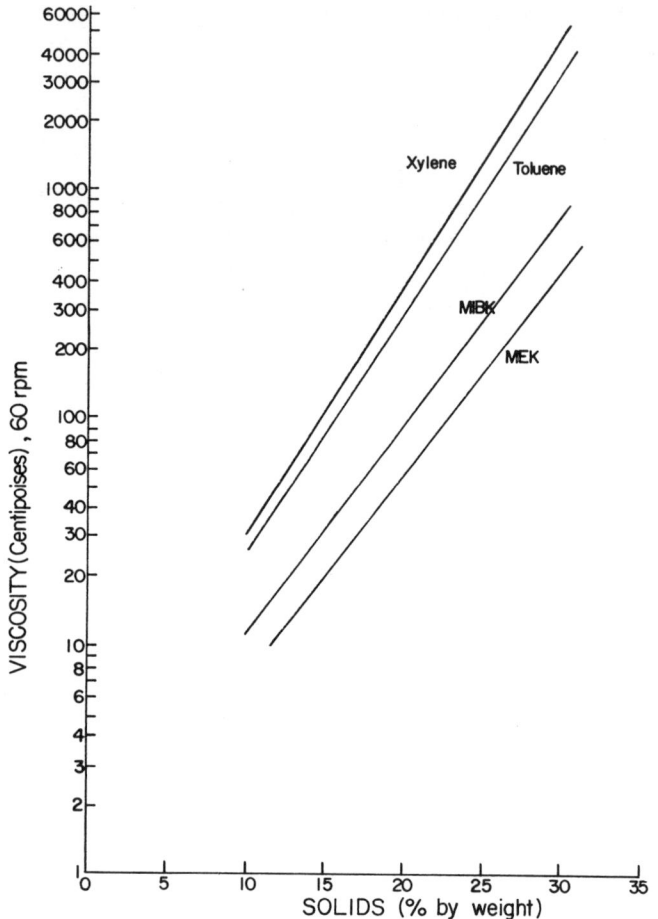

Figure 14. Solubility characteristics of a VC-TFCE copolymer (FPC 461) (Brookfield LVF viscometer)

Diluents

Diluents or organic liquids with low solvent power for the vinyl resin are frequently used to lower cost, alter evaporation rate, change rheological properties of the solution, or to achieve special characteristics such as flow out, odor, flammability, resistance to blushing, etc. A more soluble resin and a high concentration of good solvent in the system permits wider latitude in using the diluent.

Figure 22 shows the viscosity characteristics when a 30 wt % solids carboxyl containing terpolymer is used in a solvent mixture containing toluene as a diluent and strong ketone solvents; note the minimum viscosity obtained with the cyclohexanone and MIBK toluene mixtures.

Figure 23 shows the solution properties of a 20 wt % solution of the same terpolymer using xylene as the diluent. When the very soluble terpolymers are used, very low solution viscosities are obtained even though high levels of toluene or xylene are used.

High viscosity values are obtained even with the specialty copolymer. Figure 24 illustrates 30 wt % solids blend of this copolymer with various ketone/toluene mixtures. This curve can be compared with Figure 22 to visualize the effect of the polymer upon the viscosity of the solvent/diluent blends.

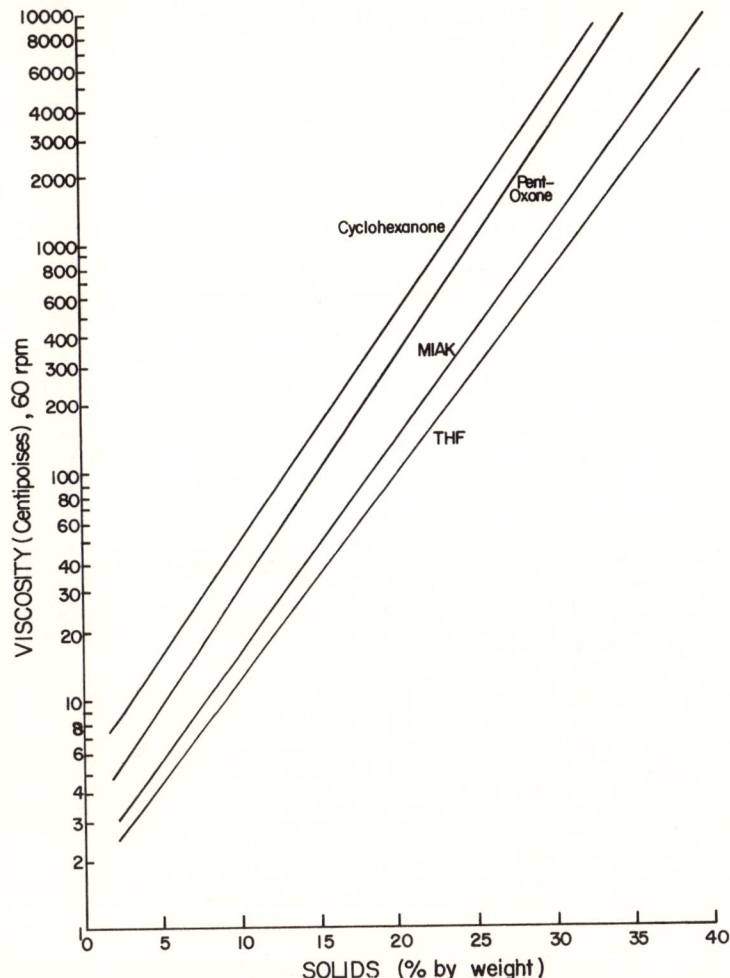

Figure 15. Solubility characteristics of a VC-TFCE copolymer (FPC 461) (Brookfield LVF viscometer)

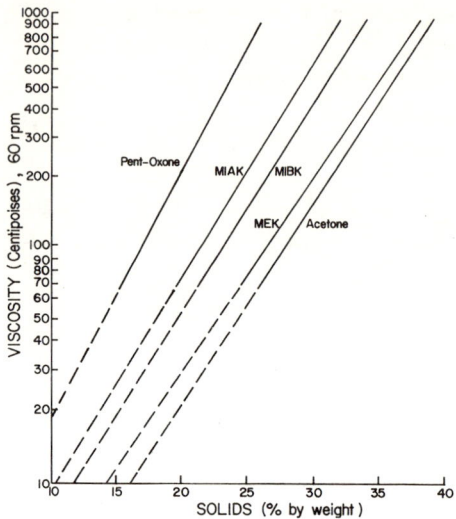

Figure 16. Solubility characteristics of a low molecular weight vinyl chloride–vinyl acetate copolymer (FPC 454) (Brookfield LVF viscometer)

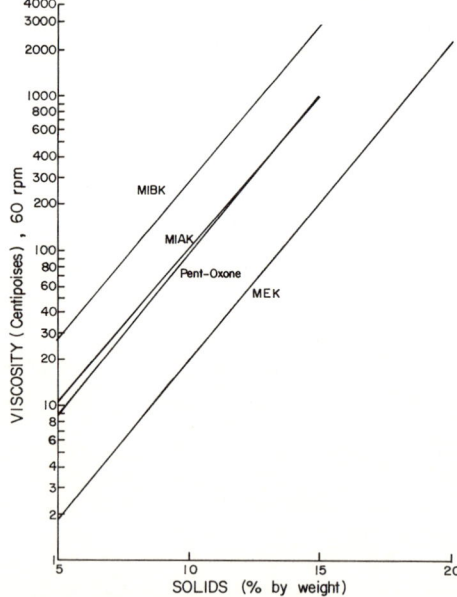

Figure 17. Solubility characteristics of a high molecular weight vinyl chloride–vinyl acetate copolymer (FPC 481) (Brookfield LVF viscometer)

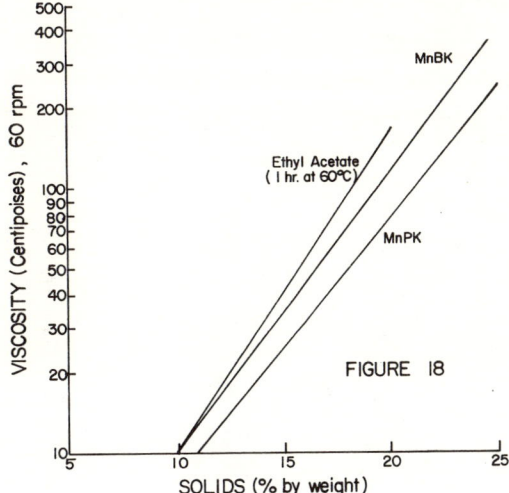

Figure 18. Solubility characteristics of a hydroxyl containing vinyl terpolymer in Rule 66-type solvents (Brookfield LVF viscometer)

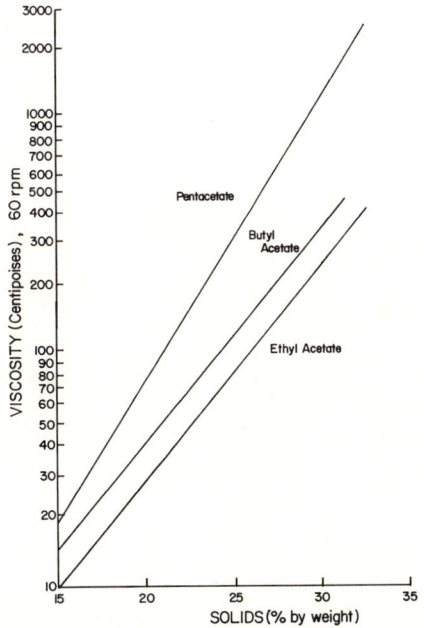

Figure 19. Vinyl terpolymer (FPC 470) in acetate solvents (Brookfield LVF viscometer)

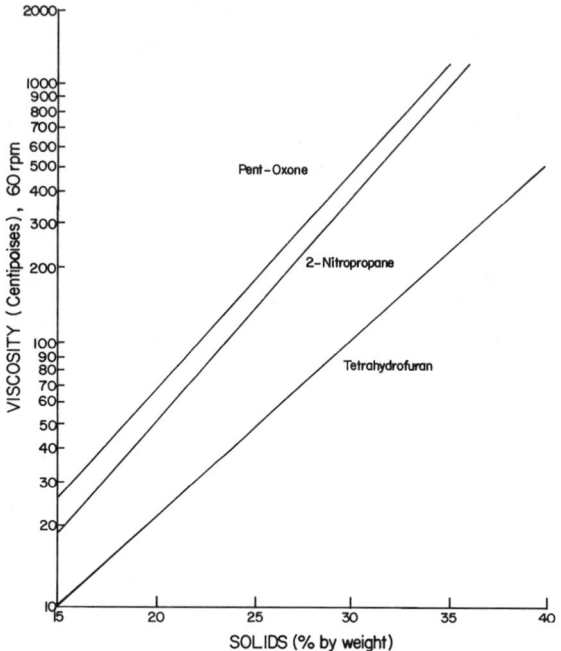

Figure 20.   Vinyl terpolymer (FPC 470) in special solvents (Brookfield LVF viscometer)

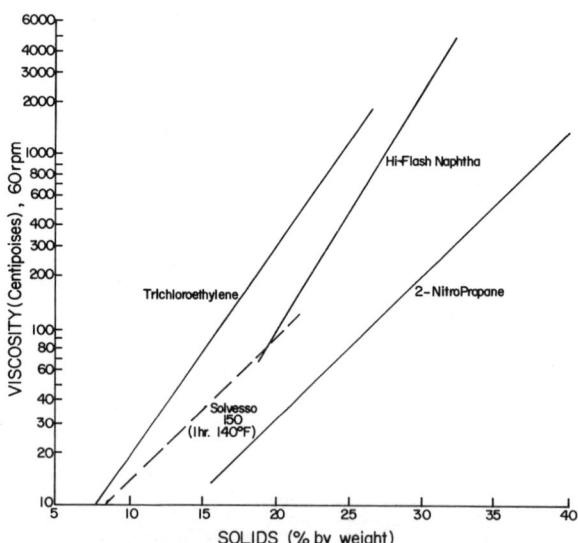

Figure 21.   Solubility characteristics of a vinyl terpolymer in special solvents (Brookfield LVF viscometer)

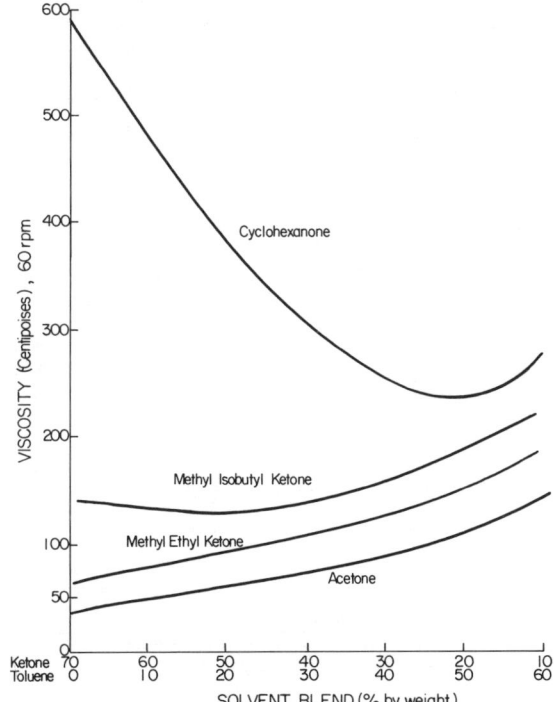

Figure 22. Solubility characteristics of a carboxyl containing vinyl terpolymer (FPC 470) in ketone/toluene blends at c0 wt % solids (Brookfield LVF viscometer)

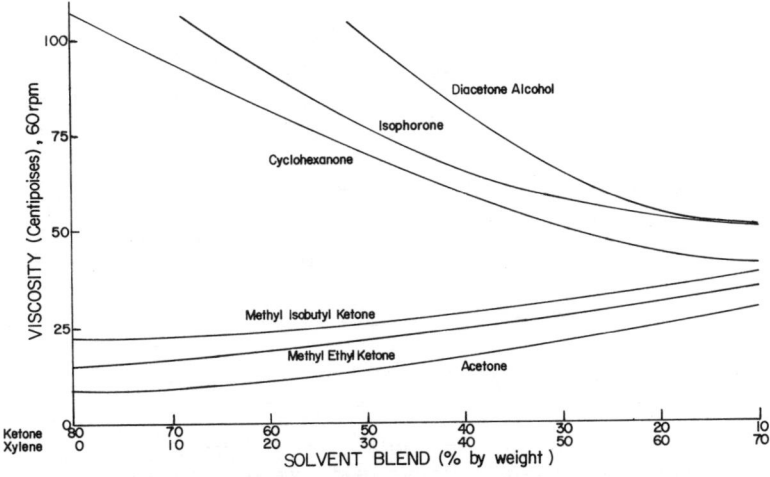

Figure 23. Solubility characteristics of a carboxyl containing vinyl terpolymer (FPC 470) in ketone/xylene blends at 20 wt % solids (Brookfield LVF viscometer)

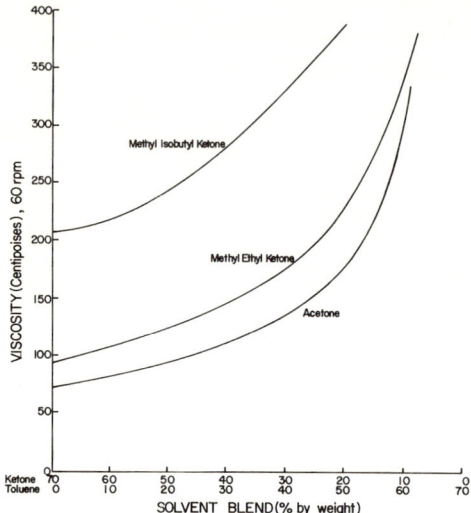

Figure 24. Solubility characteristics of a vinyl chloride specialty copolymer (FPC 471) in ketone/toluene blends at 30 wt % solids (Brookfield LVF viscometer)

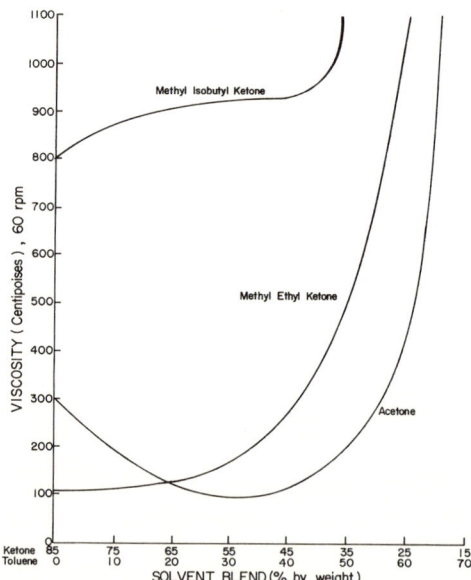

Figure 25. Solubility characteristics of a medium-high molecular weight vinyl chloride–vinyl acetate copolymer (FPC 450) in ketone/toluene blends at 15 wt % solids (Brookfield LVF viscometer)

When the vinyl chloride–vinyl acetate type copolymers are considered, tolerance for diluents drop off considerably. Figure 25 illustrates the high solution viscosities obtained when only 15 wt % of a medium high molecular weight VC–VA type copolymer is dissolved in various ketone–toluene blends.

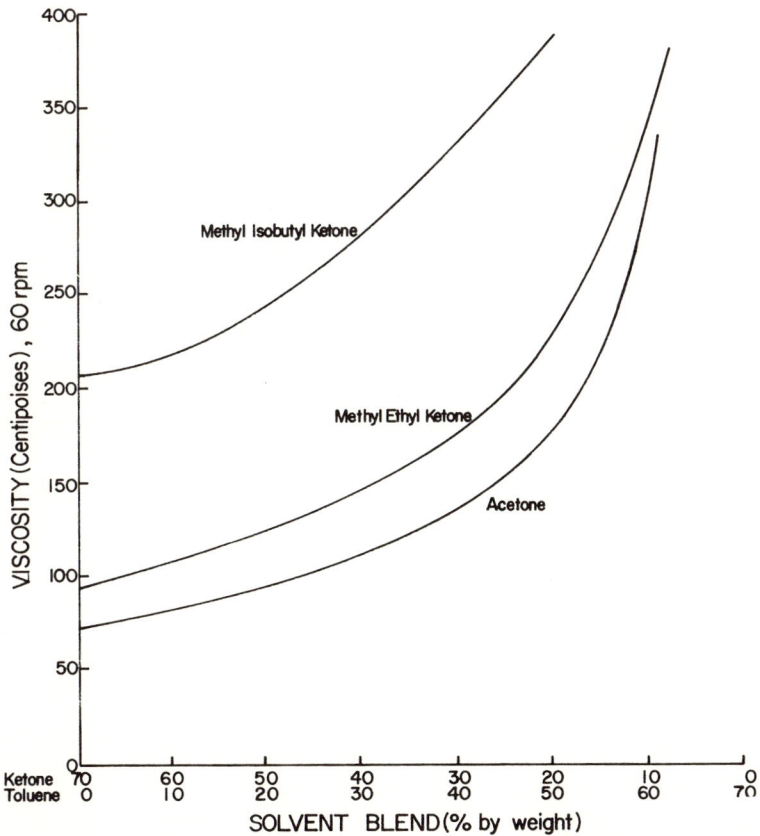

Figure 26. Solubility characteristics of a vinyl terpolymer (FPC 470) in acetate/toluene blends at 30 wt % solids (Brookfield LVF viscometer)

The acetate esters such as pentacetate, butyl acetate, and ethyl acetate give very low viscosities as diluents with the high solubility terpolymers (Figure 26). This type of polymer will also tolerate high levels of mineral spirits, high flash naphtha, VM&P naphtha, and proprietary products such as Amsco G, Solvesso 150. Some 30 wt % solid blends in MIBK are illustrated in Figure 27. Another terpolymer is soluble in high flash naphtha, trichloroethylene, and Solvesso 150 (with heating at 140°F for 1 hr). The solubility with these solvents is shown in Figure 21.

Only the lowest molecular weight vinyl chloride–vinyl acetate resins have reasonable tolerance for such poor solvents as those above. Figure 28 cites the viscosity characteristics of this type polymer in various MEK diluent blends at 25 wt % solids.

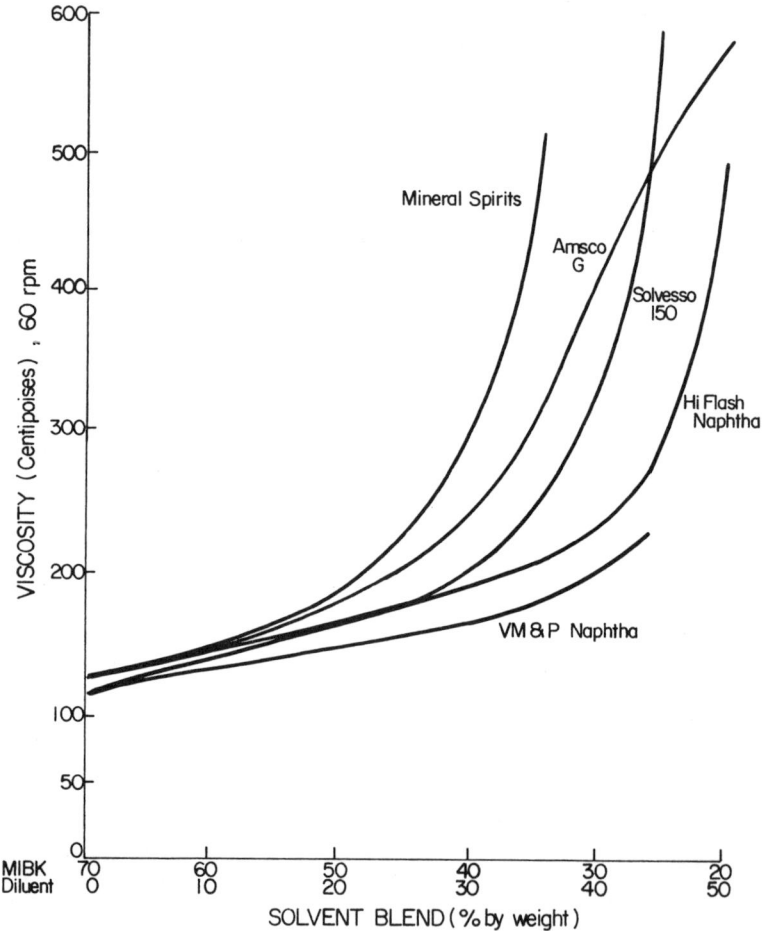

Figure 27. *Solubility characteristics of a vinyl terpolymer (FPC 470) in MIBK/diluent blends at 30 wt % solids (Brookfield LVF viscometer)*

Special Solvent Mixtures

Much has been written about Los Angeles County Rule 66 and its effect upon solvents used in vinyl solution coating formulations. The high

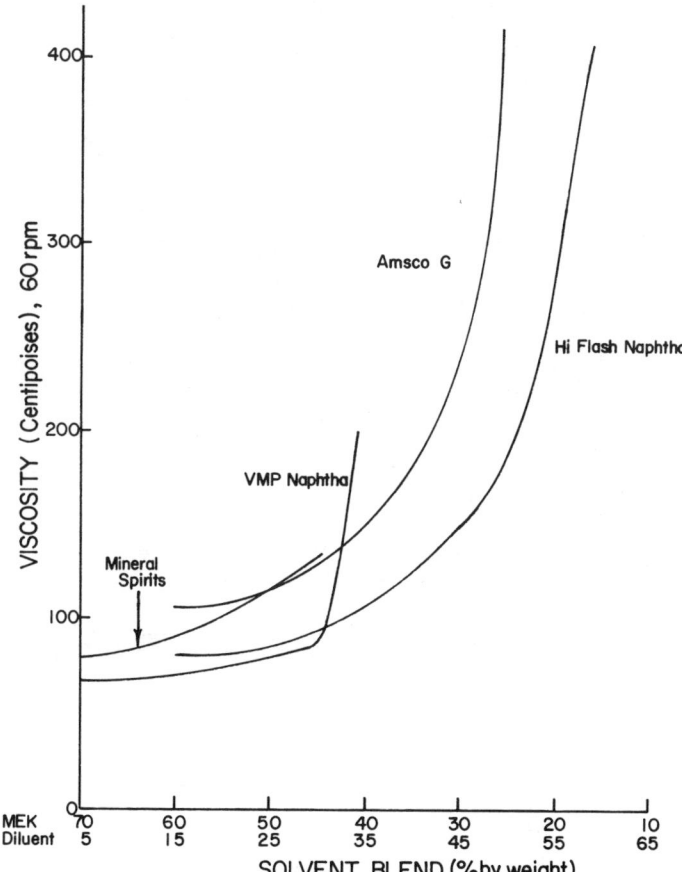

Figure 28. Solubility characteristics of a low molecular weight vinyl chloride–vinyl acetate copolymer (FPC 454) in MEK–diluent blends at 25 wt % solids (Brookfield LVF viscometer)

solubility of the vinyls available today make them well suited to meet these new solvent system requirements. Figures 29 and 30 show the solubility of various vinyl polymers in methyl n-butyl ketone and methyl n-heptyl ketone, two Rule 66-approved solvents. Figures 31 and 32 illustrate the ability of a vinyl chloride terpolymer to accommodate high levels of alcohol diluents in blends of MIBK–toluene to meet requirements of Rule 66.

The increased solubility of vinyl polymers in solvents at elevated temperatures enables very high solids to be obtained for application techniques like hot airless spraying or curtain coating. Figure 33 illustrates this property with a vinyl chloride–trifluorochloroethylene copolymer at 25° to 95°C. Since the solubility of the resin improves with

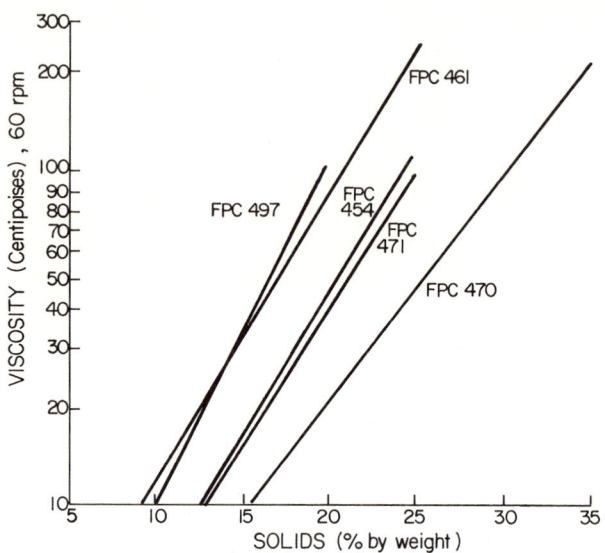

Figure 29. Solubility characteristics of vinyl polymer types in methyl n-butyl ketone (Brookfield LVF viscometer)

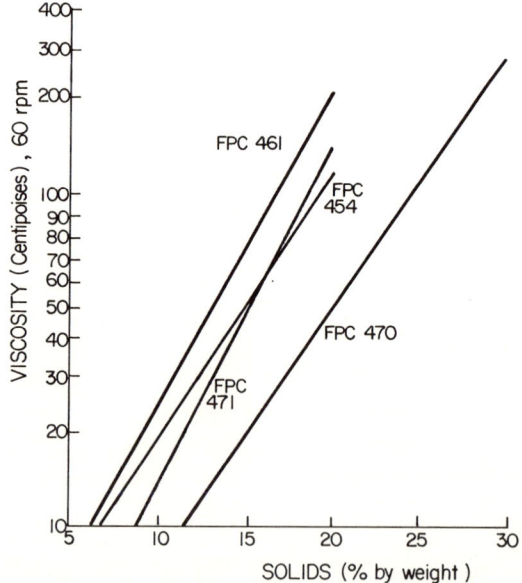

Figure 30. Solubility characteristics of vinyl polymer types in methyl n-butyl ketone (Brookfield LVF viscometer)

temperature, higher levels of diluents can be tolerated under these application techniques.

Organosols

Homopolymers and copolymers (usually vinyl acetate) of vinyl chloride are used frequently with volatile hydrocarbons where it is not desirable for the solvent (in this case really a diluent) to solvate the resin. These products are called organosols and are derived from plastisols. Plastisols and organosols have been described as follows (9):

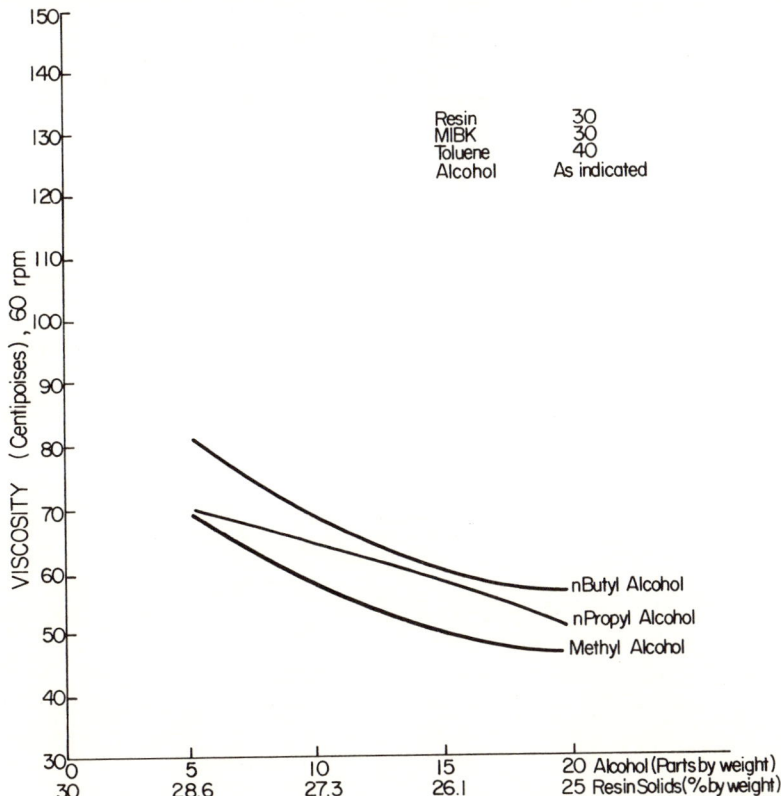

Figure 31. *Solubility characteristics of a vinyl terpolymer with alcohol diluents (Brookfield viscometer)*

Plastisols: They are defined as suspensions of homopolymers or copolymers in non-aqueous liquids. The liquids, which are normally vinyl plasticizers, are selected so that they do not solvate the polymer to any extent at room temperatures. The suspension is maintained by residual emulsifier left on the particle, and the very small particle size of the polymer itself (all pass through a 200 mesh screen). A finite quantity of plasticizer must be present in order to form the plastisol or "paste." No plasticizer. No plastisol. As with all vinyl systems, consideration may have to be given to plasticization, heat and light stabilization, pigmentation, etc. Mechanical incorporation of air into the paste or the use of chemical "blowing agents" into the formulation allows you to foam your products if desired.

Plastisols are not "air dry systems, *i.e.*, they require fusion temperatures of at least 250°F for the copolymers and 300°F for the homopolymers. Plastisols enable us to obtain very low durometer hardness, thick fused sections, and excellent chemical resistance. High durometer hardness values are difficult to obtain.

Organosols: These are plastisols which contain a volatile diluent.

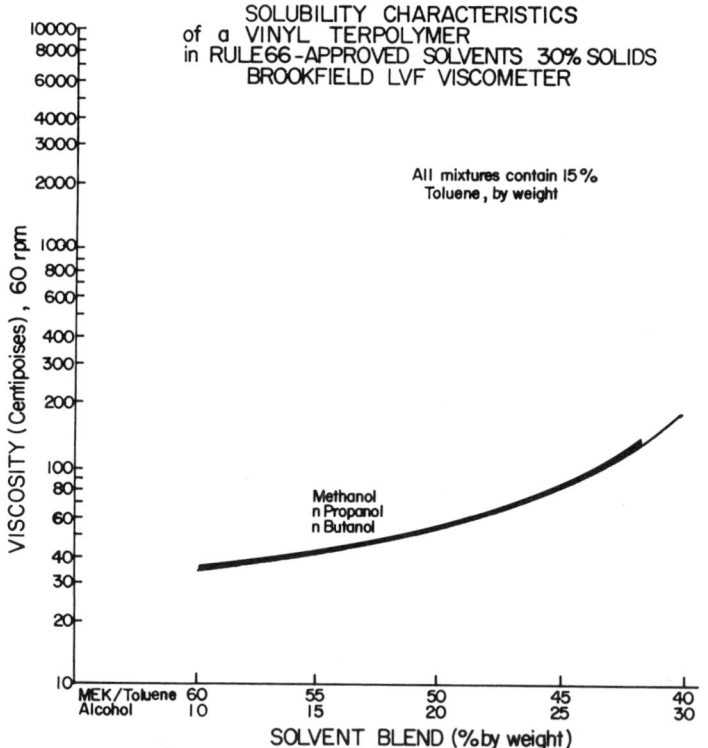

Figure 32. Solubility characteristics of a vinyl terpolymer in Rule 66-approved solvents, 30 wt % solids (Brookfield LVF viscometer)

The volatile component is usually selected from true non-solvents for PVC resins such as the aliphatic hydrocarbons. The main purpose of the volatile ingredient is to lower the viscosity of the paste by contributing more liquid to the formulation. Very low plasticizer levels are then possible and higher durometer hardness values are attainable (as compared with plastisols).

Generally, the same heat history requirements apply to the fusion or development of physicals (both organosols and plastisols). On the other hand, an organosol may require more heat input since some of the applied heat is used only for the volatilization of the diluent. The volatile component prevents you from fusing thick sections without the danger of bubble formation.

Rarely a true solvent (ketone) will be considered for organosols as an aid to gelation. Limited pot life can be expected in these situations since the resin particles will swell rapidly and give extremely high paste viscosities in short times.

This study shows the effects of diluents on viscosity stability at elevated temperatures for FPC 640 and FPC XR-2316, two popular resins

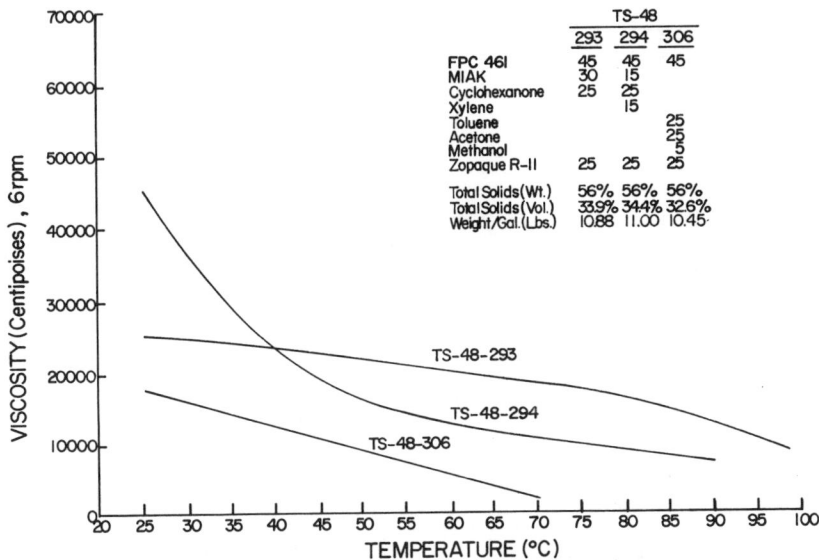

Figure 33. Low shear rate solubility characteristics of VC-TFCE copolymer (FPC 461) coatings at elevated temperatures (Brookfield LVF viscometer)

for organosol wear layer applications. The formulation used is as follows and further defined in Table V:

Resin	100
Dioctyl phthalate	35
Ferro 1776	3
Diluent	10

For comparison we used three diluents representing high, medium, and low solvating power MIBK, Solvesso 150, and Varsol 2 respectively.

Table V. Formulation

	1	2	3	4	5	6
FPC 640	100		100		100	
FPC XR–2316		100		100		100
DOP	35	35	35	35	35	35
Ferro 1776	3	3	3	3	3	3
MIBK	10	10				
Solvesso 150			10	10		
Varsol 2					10	10

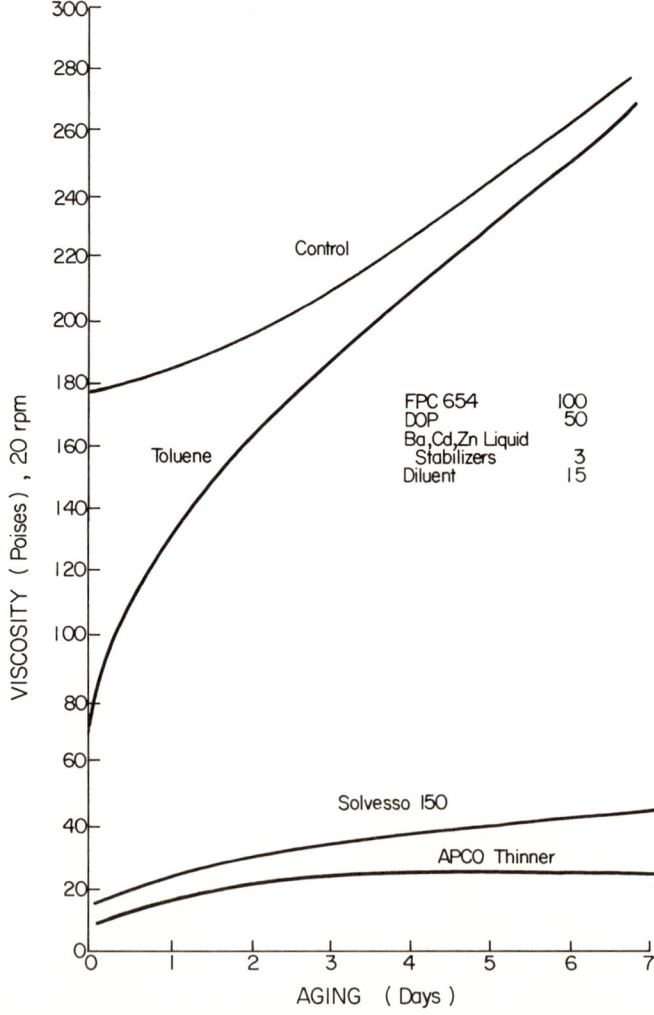

Figure 34. Organosol viscosity stability (Brookfield RVF viscometer)

The samples were prepared in half-pint cans on a Lightnin Mixer with enough shear to raise the paste temperature to 100°F. Brookfield viscosities at 2/20 rpm were obtained at 100°F immediately after mixing 2 hours, 1, 4, 5, 7, 21, and 28 days (Table VI). The diluent with the least solvating power, Varsol 2 gave the best viscosity stability with both FPC 640 and FPC XR-2316.

Table VI. Viscosity Stability at 100°F[a]

2/20 RPM Brookfield, P	1	2	3	4	5	6
Initial	4600	6300	70	250	55	90
	1150	TVTM[b]	26	79	21	30
2 Hs	TVTM	TVTM	350	1400	130	180
			95	290	39	50
1 Day	TVTM	TVTM	3500	14,720	530	905
			782	TVTM	130	198
4 Day	TVTM	TVTM	TVTM	TVTM	1200	2400
					270	530
5 Day	TVTM	TVTM	TVTM	TVTM	1280	2940
					318	592
7 Day	TVTM	TVTM	TVTM	TVTM	1500	3200
					300	660
21 Day	TVTM	TVTM	TVTM	TVTM	2920	7700
					612	TVTM
28 Day	TVTM	TVTM	TVTM	TVTM	3500	TVTM
					730	

[a] See table V for formulation.
[b] TVTM, too viscous to measure.

Table VII. Tensile Development[a]

10 Min at 150°C	1	2	3	4	5	6
Tensile, psi	[b]	[b]	2070	1950	1830	1810
Elongation, %			130	110	120	110
100% modulus, psig			1640	1740	1590	1720
10 Min at 180°C						
Tensile, psig	[b]	[b]	3380	3200	3260	3360
Elongation, %			280	270	280	270
100% modulus, psig			1790	1790	1770	1990

[a] See table V for formulation.
[b] Paste too viscous to make films.

Table VII shows tensile development at 150° and 180°C for Solvesso 150 and Varsol 2. Pastes produced with MIBK as diluent were too viscous even at the early stages to make up films for this particular study.

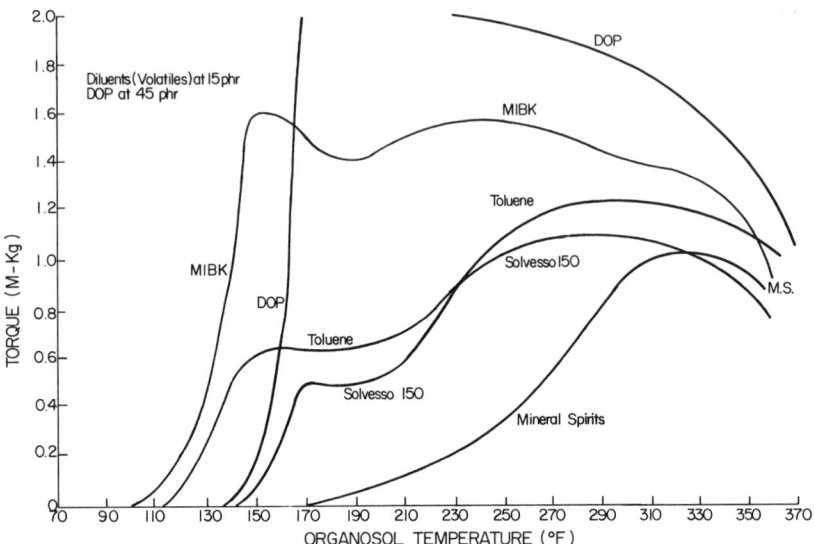

Figure 35. *Torque rheometer evaluation of organosol diluents*

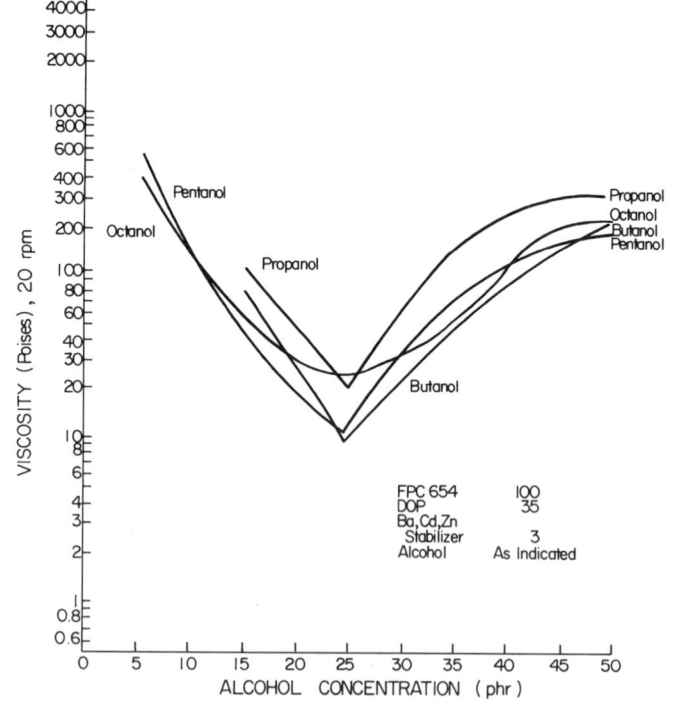

Figure 36. *Organosol viscosity study of alcohols (Brookfield RVF viscometer)*

In summary, slower solvating type diluents tend to have much better viscosity stability and generally lower cost than the high solvators like the aliphatic ketones. These concepts, coupled with the fact that the slow solvators are generally less hazardous, more than compensate for the small loss in tensile development at the lower temperatures.

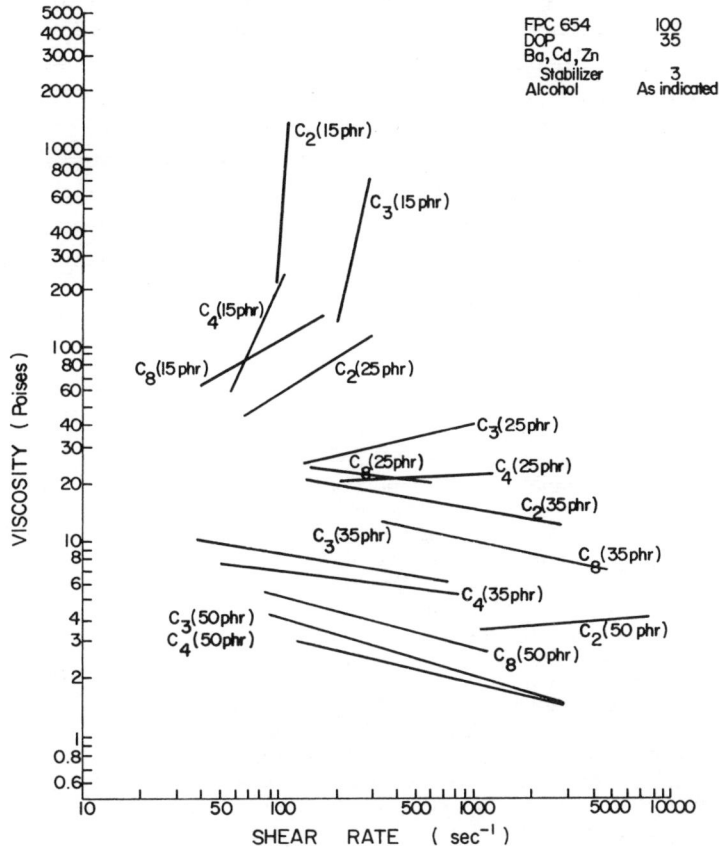

Figure 37. *Organosol viscosity study of alcohols (Severs extrusion rheometer)*

Figure 34 shows the viscosity stability of a conventional organosol using a general purpose dispersion resin in conjunction with toluene, and two commercial mixed diluents (Solvesso 150 and APCO thinner). All of these diluents lower the initial paste viscosity, but the viscosity stability of the organosol can vary widely depending on the chemical composition of the volatile component. The chemical composition—i.e., solvent power—of the diluent will not only affect viscosity stability but also gelation and fusion rates. This is shown graphically in Figure 35 which uses a torque-rheometer (*10*) to measure viscosity from the paste (fluid) form to the fused (hot melt) form. This instrument records viscosity in units of torque (meter-kilograms). The toluene and Solvesso 150, while relatively inert at room temperature, become much stronger solvents when temperatures are elevated.

Ogorzalek (*11*) has studied in detail the effect of shear rate on the viscosity of organosols. His data for low shear (Brookfield) and high shear (Extrusion Rheometer) viscosity characteristics using alcohols as diluents are shown in Figures 36 and 37.

With organosols we are formulating to resist solvation of the resin by the volatile component. We are trying to take advantage of its fluid properties only. This is directly opposite to our formulating approach with solution or soluble vinyl resins, *i.e.,* incomplete solvation of the resin by the solvent will result in erratic viscosity and lowered physical properties of the final film.

Literature Cited

1. Products Data, FPC 470 or FPC 477, Firestone Plastics Co., Box 699, Pottstown, Pa. 19464.
2. Reichard, R., *Paint Varn. Prod.* (1970) **60** (10), 63.
3. Park, R. A., *Viscometers*, Firestone Plastics Co., **19464**, 1.
4. SPI-VD-T-19 *Procedure To Determine A Plastisol's Apparent Viscosity Utilizing A Torque Rheometer (Brabender Plasti-corder) At Constant Mixer Temperature*, The Society of The Plastics Industry, Vinyl Dispersions Division, New York, 1968.
5. Park, R. A., Klinger, J. L., *Amer. Paint J.* (1970) **55** (13), 87.
6. Park, R. A., *Paint Varn. Prod.* (1965) **53** (4), 35.
7. Doolittle, A. K., "The Technology of Solvents and Plasticizers," John Wiley and Sons, New York (1954).
8. Park, R. A., *Paint Varn. Prod.* (1965) **55** (6), 51.
9. Park, R. A., *Plast. Technol.* (1970) **16** (6), 46.
10. Park, R. A., *Paint Varn. Prod.* (1965) **55** (7), 45.
11. Ogorzalek, J. M., "The Effect of Diluents Upon Stir-In Organosol Rheological and Fusion Characteristics," presented at 21st ANTEC, Soc. of Plast. Eng., Boston, 1965.

RECEIVED March 10, 1972.

INDEX

INDEX

A

Abstraction, hydrogen	107
Acetates	107
Acid theory, Lewis	6
Acrylic resin	148
dilution curves for an	156
solutions, rupture voltages of	157
Activity	
coefficients, limiting group	14
curves, master	19
resin	22
Adhesives	138
hot melt	139
Aerosol	
atmospheric	81
formation, influence of SO_2 on photochemical	92
Aggregation-disaggregation equilibria	32
Air dry systems	211
Alcohols	107
Aliphatic/alcohol tolerance	116
Alkanes	107
Alkyd	
resins	
group compositions of short oil	34
properties of short oil	34
viscosities of short oil	35
trimellitic	149
Amides	108
Amperage plot, time-	144
Analysis (ERA), evaporation rate	119
Analytical solution of groups (ASOG)	11, 12, 32
Analytical techniques	97
Application/performance properties	118
Applications in coatings and inks, solvent	113
Aqueous solution preparation, typical	161
Aromatics	105
(ASOG), analytical solution of groups	32
ASOG matrix development	13
Atmospheric	
aerosol	81
reactions of solvents, photochemical smog and the	95
reactivities, scale of	95

B

Background air used in the experiments	72

Bar, New York Paint Committee leveling	122
Bisphenol-A, (BPA)	176
Bond accepting index, net hydrogen	4
Bonding forces, hydrogen-	48
Bonding indexes (γ), hydrogen	114
(BPA), bisphenol-A	176

C

Calculated and experimental TOC flash points	60
Calculating concentrated polymer solution viscosities	31
Chamber, investigations in the steel	98
Chamber conditions, irradiation	97
Characteristics of vinyl chloride homopolymers, copolymers and terpolymers, solubility	186
Chart, viscosity conversion	121
Chemical parameters of solvents, physical	176
Chemical reactions leading to photochemical smog	103
Chloride homopolymers, copolymers and terpolymers, solubility characteristics of vinyl	186
Classifications, molecular weight	191
Classification of solvents, Pimentel and McCellan	4
Clustering theories, molecular	7
Coatings	
film substrate attack by solvents, inks, and	127, 129
and inks	139
solvent applications in	113
reformulating solvent mixtures in epoxy resin	175
Coefficients, limiting group activity	14
Coefficients, for a steel chamber correlation	104
Combinations, solubilities for heptane	172
Combinations, solubilities for methanol	172
Combustion, heat of	59
Comparison of	
oxidant–maximum rankings	84
solvent data	87
TOC flash points	67
Component flash point, pure	59
Composition	
of mixed solvents	42
during mixed solvent evaporation from resin solutions	11

221

Composition *(Continued)*
of short oil alkyd resins, group .. 34
Computer, solvent selection by 48
Concentrated polystyrene solutions. 42
Concentrated resin solutions 34
Concentration, critical 36
Conversion chart, viscosity 121
Copolymers and terpolymers,
solubility characteristics of
vinyl chloride homopolymers.. 186
Copolymers, vinyl chloride 187
Correlation coefficients for a
steel chamber 104
Correlations between various
measures of photochemical
smog 100
Cosolvent on wet film resistance
of the epoxy ester, effect of... 148
Cost determination, optimum
viscosity 118
Coulombic efficiencies for an epoxy
ester 153
Coulombic efficiency 144
Criteria for photochemical reactivity 72
Criteria, solvency 51
Critical concentration 36
Crosslinking 186
Cup (TOC) flash point, Tagliabue
open 57
Curves
for an acrylic resin, dilution 156
for the epoxy ester, dilution 149
master activity 19

D

δ, solubility parameter 2, 168
Data, comparison of solvent 87
Data, formulations and 127
Development, ASOG matrix 13
Diagrams, ternary solubility 178
Diluents 200
Dilution curves for an acrylic resin. 156
Dilution curves for the epoxy ester. 149
Dipole
-induced dipole forces,
permanent 49
moment force 169
-permanent dipole forces,
permanent 48
Disappearance, rates of
hydrocarbon 100
Dispersion forces 169
Dispersion, pigment 9
Dosage rankings, oxidant 84
Dry times of polyamide inks 119
Durometer hardness values 212

Effect *(Continued)*
of voltage and solids on film
thickness 151, 154
Effective intrinsic viscosity 33
Efficiency, coulombic 144
Efficiencies for an epoxy ester,
coulombic 153
Electrodeposition procedure 164
Electrodeposition resins 141
Electrostatic spraying, polar
solvents of 126
Energy, interfacial free 8
Epoxy
ester
coulombic efficiencies for an.. 153
dilution curves for the 149
effect of cosolvent on wet
film resistance of the 148
solutions, rupture voltages of . 150
(FTE), flexibized thermoplastic. 182
resin
coatings, reformulating solvent
mixtures in 175
ester 148
(TE), thermoplastic 176
Equation
Martin 33
Mark-Houwink 33
Van Laar 67
Equilibria, aggregation-
disaggregation 32
Equilibria, liquid-liquid 22
Equivalents, toluene 98
(ERA), evaporation rate analysis.. 119
Ester
coulombic efficiencies for an
epoxy 153
dilution curves for the epoxy ... 149
effect of cosolvent on wet film
resistance of the epoxy 148
epoxy resin 148
solutions, rupture voltages of
epoxy 150
Ethers 107
Evaporation
rate 52
analysis (ERA) 119
from resin solutions, composition
during mixed solvent 11
tests 23
times of solvents 122
Experimental reactivity 71
Experimental TOC flash points,
calculated and 60
Eye-irritation rankings 86
Eye-irritation response 74

E

Effect
of cosolvent on wet film resistance
of the epoxy ester 148
of initial nitrogen dioxide 86

F

Film
penetration test 129
resistance of the epoxy ester,
effect of cosolvent on wet .. 148

INDEX

Film *(Continued)*
- resistance 148
- wet 144
- solvent retention in vinyl 123
- substrate attack by solvents, inks, and coatings 129
- thickness 145
  - effect of voltage and solids on. 151
- weight 145

Flash point
- calculated and experimental TOC for mixed solvents, predicting TOC 60
- TOC 59
- pure component 59
- solvent blend TOC 61
- for solvents mixtures, predicting. 56
- Tagliabue open cup (TOC) 57

Flexibilized thermoplastic epoxy, (FTE) 182
Flory-Huggins theory 6
Flow/wetting properties 127
Fluids, non-Newtonian 193
Force, dipole moment 169

Forces
- dispersion 169
- hydrogen-bonding 48
- London 48
- permanent dipole-induced dipole 49
- permanent dipole-permanent dipole 48

Formation, rates of nitrogen dioxide 99
Formulations and data 127
Free energy, interfacial 8
(FTE), flexibilized thermoplastic epoxy 182

G

γ, hydrogen bonding indexes 114
Gelation 217
Group compositions of short oil alkyd resins 34
Groups (ASOG) method, analytical solution of 11, 12

H

Hansen's three-component parameter 3
Hardness values, durometer 212
Heat of combustion 59
Heat of vaporization 2
Heptane combinations, solubilities for 172
Hildebrand solubility parameter method 12
Hildebrand theory of solubility ... 1
Homopolymers 187
Homopolymers, copolymers and terpolymers, solubility characteristics of vinyl chloride ... 186
Hot melt adhesives 139

Hydrocarbon
- disappearance, rates of 100
- resins
  - low molecular weight 131
  - solvent systems for 131
Hydrodynamic molecular weight .. 32

Hydrogen
- abstraction 107
- bond accepting index, net 4
- bond
  - classes of solvents 2
  - quantifying the 4
- bonding forces 48
- bonding indexes (γ) 114

Hypothetical solids-to-solvent Rule 66 situation 91

I

Improvement, solubility parameter. 53
Indene polymers 135
Indentation (M), rate of 137
Index, net hydrogen bond accepting 4
Indexes (γ), hydrogen bonding ... 114
Induced dipole forces, permanent dipole- 49
Influence of SO_2 on photochemical aerosol formation 92
Initial nitrogen dioxide, effect of .. 86

Inks
- and coatings, film substrate attack by solvents, ... 129
- dry times of polyamide 119
- polyamide 122
- solvent applications in coatings and 113

Interaction, segment-segment 32
Interactions of resins, segmental .. 32
Interfacial free energy 8
Investigations in the steel chamber. 98
Irradiation chamber conditions ... 97
Irritation rankings, eye 86
Isotherms, resin 19

K

Ketones 109

L

Leveling 11
Leveling bar, New York Paint Committee 122
Lewis acid theory 6
Limiting group activity coefficients 14
Liquid-liquid equilibria 22
London forces 48
Los Angeles County Rule 66 .... 5, 50, 70, 95, 125

M

Maps, solubility 175
Mark-Houwink equation 33

224   SOLVENTS THEORY AND PRACTICE

Martin equation	33
Master activity curves	19
Matrix development, ASOG	13
Maximum operating voltage (MOV)	144
Maximum rankings, comparison of oxidant–	84
Mean vapor molecular weight	59
Measures of photochemical smog, correlations between various	101
Measurements, viscosity	193
Melt adhesives, hot	139
Methanol combinations, solubilities for	172
Method, analytical solutions of groups (ASOG)	11, 12
Method, Hildebrand solubility parameter	12
Miscibility	11
Mixed solvents, composition of	42
Mixtures in epoxy resin coatings, reformulating solvent	175
Molecular clustering theories	7
Molecular weight	
classifications	191
hydrocarbon resin, low	131
hydrodynamic	32
mean vapor	59
Moment force, dipole	169
(MOV), maximum operating voltage	144

N

Neat solvent viscosity	176
Net hydrogen bond accepting index	4
New York Paint Committee leveling bar	122
Nitric oxide (NO_2 t-max), photo-oxidation	75
Nitrogen dioxide	
effect of initial	86
formation, rates of	99
Non-Newtonian fluids	193

O

Olefins	105
Open Cup (TOC) flash point, Tagliabue	57
Operating voltage (MOV), maximum	144
Optimum viscosity/cost determination	118
Organic solvents, photochemical smog reactivity of	70
Organosols	209
Oxidant	
dosage rankings	84
–maximum rankings, comparison of	84
production	74

P

Parameter	
Hansen's three-component	3
improvement, solubility	53
method, Hildebrand solubility	12
numbers (δ), solubility	114
plot, solubility	50
(δ), solubility	2, 133, 168, 175
Parameters	
of resins	132
of solvents	132
physical chemical	176
for specific solvents, solubility	170
Penetration test, film	129
Performance properties, application	118
Permanent dipole-induced dipole forces	49
Permanent dipole-permanent dipole forces	48
Phase separation	134
Phase splits	22
Photochemical	
aerosol formation, influence of SO_2 on	92
reactivity, criteria for	72
smog	
and the atmospheric reactions of solvents	95
chemical reactions leading to	103
correlations between various measures of	101
production	96
reactivity of organic solvents	70
Photooxidation of nitric oxide (NO_2 t-max)	75
Photooxidation of trichloroethylene	85
Physical chemical parameters of solvents	176
Pigment dispersion	9
Pimentel and McClellan classification of solvents	4
Plasticizers	131
Plastisols	211
Plot, time-amperage	144
Polar solvents for electrostatic spraying	126
Polyamide inks	119, 122
Polyamide resin solubility	168, 169, 173
Polyamide resin specifications	170
Polymer solution viscosities, calculating concentrated	31
Polymers, indene	135
Polystyrene solutions, concentrated	42
Poly(vinyl chloride) (PVC)	124, 187
Power, throwing	144
Predicting flash points for solvent mixtures	56
Predicting TOC flash points for mixed solvents	58
Preparation, typical aqueous solution	161

INDEX

Prigogine theory 5
Procedure, electrodeposition 164
Production, oxidant 74
Production, photochemical smog .. 96
Profile, smog 71
Properties
   application performance 118
   flow/wetting 127
   of short oil alkyd resins 34
Pure component flash point 59

Q

Quantifying the hydrogen bond .. 4

R

Rankings
   comparison of oxidant–maximum 84
   eye-irritation 74, 86
   oxidant dosage 84
Rate
   analysis (ERA), evaporation ... 119
   evaporation 52
   of hydrocarbon disappearance .. 100
   of indentation, (M) 137
   of nitrogen dioxide formation .. 99
Reactions leading to photochemical smog, chemical 103
Reactions of solvents, photochemical smog and the atmospheric 95
Reactivities
   scale of atmospheric 95
   solvent 106
Reactivity
   criteria for photochemical 72
   experimental 71
   of organic solvents, photochemical smog 70
   scale, Rule 66 72
Reformulating solvent mixtures in epoxy resin coatings 175
Reformulation, solvent 91
Resin 142
   acrylic 148
   activity 22
   coatings, reformulating solvent mixtures in epoxy 175
   dilution curves for an acrylic ... 156
   ester, epoxy 148
   isotherms 19
   low molecular weight hydrocarbon 131
   solubility, polyamide 173, 169, 173
   solution viscosity 12
   solutions
      compositions during mixed solvent evaporation from . 11
      concentrated 34
      rupture voltages of acrylic.... 157
   specifications, polyamide 170
   (TE), thermoplastic epoxy 176
Resins
   electrodeposition 141
   group compositions of short oil alkyd 34
   parameters of 132
   properties of short oil alkyd 34
   segmental interactions of 32
   solvent systems for hydrocarbon. 131
   viscosities of short oil alkyd.... 35
Resistance of the epoxy ester, effect of cosolvent on wet film 148
Resistance, wet film 144
Rubber compounds 138
Rule 66
   Los Angeles County..5, 50, 70, 95, 125
   reactivity scale 72
   situation, hypothetical solids-to-solvent 93
Rupture voltages 144
   of acrylic resin solutions 157
   of epoxy ester solutions 150

S

S position 147
Sample ternary diagrams 178
Scale of atmospheric reactivities.. 95
Segmental interactions of resins ... 32
Segment-segment interaction 32
Selection by computer, solvent 48
Separation, phase 134
Short oil alkyd resins
   group compositions of 34
   properties of 34
   viscosities of 35
Smog
   and the atmospheric reactions of solvents, photochemical ... 95
   correlations between various measures of photochemical. 101
   production, photochemical 96
   profile 71
   reactivity 71
      of organic solvents, photochemical 70
SO_2 on photochemical aerosol formation, influence of 92
Solids on film thickness, effect of voltage and 151
Solids-to-solvent Rule 66 situation, hypothetical 91
Solubilities for heptane combinations 172
Solubilities for methanol combinations 172
Solubility
   characteristics of vinyl chloride homopolymers, copolymers, and terpolymers 186
   diagrams, ternary 178
   Hildebrand theory of 1
   maps 175

Solubility *(Continued)*
　parameter
　　δ 2, 132, 168, 175
　　improvement 53
　　method, Hildebrand 12
　　numbers (δ) 114
　　plot 50
　　for specific solvents 170
　polyamide resin 168, 169, 173
　viscosity 114
Solution
　preparation, typical aqueous ... 161
　stability 146
　viscosity, resin 12
Solutions
　compositions during mixed
　　solvents evaporation
　　from resin 11
　concentrated polystyrene 42
　concentrated resin 34
　of groups method,
　　analytical 11, 12
　rupture voltages of acrylic resin. 157
　rupture voltages of epoxy ester.. 150
Solvation-desolvation equilibria ... 32
Solvency criteria 51
Solvent
　applications in coatings and inks. 113
　blend TOC flash points 61
　data, comparison of 87
　evaporation from resin solutions,
　　compositions during mixed . 11
　mixtures in epoxy resin coatings
　　reformulating 175
　mixtures, predicting flash
　　points for 56
　reactivities 106
　reformulation 91
　retention in vinyl films 123
　Rule 66 situation, hypothetical
　　solids-to- 91
　science, trends in 1
　selection by computer 48
　systems for hydrocarbon resins.. 131
　technology, trends in 1
　viscosity, neat 176
Solvents
　composition of mixed 42
　for electrostatic spraying, polar.. 126
　evaporation times of 122
　inks, and coatings, film substrate
　　attack by 129
　parameters of 132
　photochemical smog and the
　　atmospheric reaction of 95
　photochemical smog reactivity
　　of organic 70
　physical chemical parameters of. 176
　Pimentel and McClellan
　　classification of 4
　predicting TOC flash points
　　for mixed 58
　solubility parameters for specific. 170
Specifications, polyamide resin ... 170

Spraying, polar solvents for
　electrotatic 126
Stability, solution 146
Steel chamber, investigations in the 98
Systems for hydrocarbon resins,
　solvent 131

T

Tagliabue closed cup (TCC)
　flash point 57
Tagliabue open cup (TOC)
　flash point 57
(TE), thermoplastic epoxy resin .. 176
Ternary diagrams, sample 178
Ternary solubility diagrams 178
Terpolymers, solubility characteristics of vinyl chloride
　homopolymers, copolymers,
　and 186
Terpolymers of vinyl chloride 189
Test, film penetration 129
Tests, evaporation 130
Tetrahydrofuran, (THF) 187
Theory
　Flory-Huggins 6
　Lewis acid 6
　Prigogine 5
　of solubility, Hildebrand 1
Theories, molecular clustering 7
Thermoplastic epoxy, (FTE),
　flexibilized 182
Thermoplastic epoxy resin, (TE).. 176
(THF), tetrahydrofuran 187
Thickness, effect of voltage and
　solids on film 151
Thickness, film 145
Throwing power 144
Time-amperage plot 144
Times of solvents, evaporation ... 122
TOC flash points
　Tagliabue open cup 57
　calculated and experimental 60
　for mixed solvents, predicting .. 58
　solvent blend 61
Tolerance, aliphatic/alcohol 116
Toluene equivalents 98
Torque-rheometer 191
Trichloroethylene 88
Trifluorochloroethylene, vinyl
　chloride– 188
Trimellitic alkyd 149
Trends in solvent science 1
Trends in solvent technology 1
Typical aqueous solution
　preparation 161

V

Values, durometer hardness 212
Van Laar equations 67
Vapor molecular weight, mean ... 59

Vaporization, heat of	2
Vinyl	
chloride	
copolymer	187
homopolymers, copolymers, and terepolymers, solubility characteristics of	186
terpolymers of	189
trifluorochloroethylene	188
films, solvent retention in	123
Viscosities, calculating concentrated polymer solution	31
Viscosities of short oil alkyd resins	35
Viscosity	
conversion chart	121
/cost determination, optimum	118
effective intrinsic	33
measurements	193
neat solvent	176
resin solution	12
solubility	114
Voltage	
(MOV), maximum operating	144
rupture	144
and solids on film thickness, effect of	151
Voltages of acrylic resin solutions, rupture	157
Voltages of epoxy ester solutions, rupture	150

W

Weight	
classifications, molecular	191
film	145
resistance	144
mean vapor molecular	59
Wet film resistance of the epoxy ester, effect of cosolvent on	148
Wetting properties, flow	127